Pre-Accident Investigations

To everyone who has ever asked "how" instead of "why."

Pre-Accident Investigations

An Introduction to Organizational Safety

TODD CONKLIN
Los Alamos National Laboratory, USA

CRC Press
Taylor & Francis Group
Boca Raton London New York

CRC Press is an imprint of the
Taylor & Francis Group, an **informa** business

CRC Press
Taylor & Francis Group
6000 Broken Sound Parkway NW, Suite 300
Boca Raton, FL 33487-2742

© 2012 by Todd Conklin
CRC Press is an imprint of Taylor & Francis Group, an Informa business

No claim to original U.S. Government works

Printed on acid-free paper
Version Date: 20160226

International Standard Book Number-13: 978-1-4094-4783-2 (Hardback) 978-1-4094-4782-5 (Paperback)

Visit the Taylor & Francis Web site at
http://www.taylorandfrancis.com

and the CRC Press Web site at
http://www.crcpress.com

Contents

List of Figures

List of Abbreviations

ATIS	Automated Traffic Information System "Sierra"
CQD	Antiquated nautical "attention all stations" distress or danger call
DFEO	Deviation From Expected Outcome
HAZCOM	Hazard Communication
HR	Human Resources
JHA	Job Hazard Analysis
JSA	Job Safety Analysis
LANL	Los Alamos National Labortatory
LOTO	Lock Out Tag Out
MPH	Miles Per Hour
MYG	Titanic call letters
OSHA	Occupational Safety and Health Administration
RCA	Root Cause Analysis
SOS	Nautical distress call
TOGA	Flight deck intercom

Foreword
A Context Setting Discussion

Let's start with a story. Humans have an unusually strong bias towards learning through storytelling. I intend to use this same bias to begin this book. In many ways, this story covers everything that we will discuss in this book: the presence of normal operational information in every event; the belief that had we had access to the right knowledge before a failure happened we would have most certainly avoided the failure; the clear absence of failure identification before the failure started. Read this story, and think about how these ideas could apply to your organization. After all, with a couple of little changes, this story could be your organization's story.

A colleague of mine who is a noted high explosive researcher told me this story. In many ways, this story represents the central theme of this book. By retelling this story to you, I am both setting the stage, and engaging your experience in order to make the ideas and concepts more successful and practical for you.

In the history of studying high explosives, an important expert and early father of high explosive research and safety learned a critical lesson. This expert had done hundreds of high risk, highly technical experiments with all sorts of things that blow up. With all this experience, our expert not only became technically famous, but he also became an expert in how to perform this type of work safely.

This expert knew that every time an experiment went wrong, a line of fellow researchers would be at his office door the next morning wanting to tell him why they thought the failed experiment had failed. Sometimes the line went all the way down the hall. Each researcher in this line had a thoughtful idea or two on "what went wrong," and how they could help prevent this problem from happening the next time an experiment was to be completed.

Every failure was followed by a formal technical meeting to discuss what happened in the failure. The researcher noted that his fellow scientists had a particularly strong need to have some type of post-mortem discussion about the failure after the failure. Almost like a cleansing ritual. These discussions were always intriguing. Many times the information volunteered during these informal post-mortems would have made a difference to the overall success of the failed experiment. These researchers all seemed to have some pretty clear ideas and observations.

This expert thought, "If all this knowledge exists after a failure—could at least some of it exist before a failure?" He realized that though some information was clearly only available in retrospect, some must be available before the next experiment was to be fired. Knowing this information before an experiment is done seemed like a very good idea. In fact, responding to this information before the failure happened would be much more efficient and effective, cheaper and faster, and much better for the future of high explosive science.

And that's the moment when the idea of a "premortem" for high explosive experimentation was born. It is being the Crime Scene Investigator before the crime, and identifying the possible "hows" before the event.

Having a premortem meeting, a meeting where you ask smart, experienced people what could go wrong before it does go wrong, provides a new set of data about a failure that has yet to happen. Knowing this new information allowed the researchers to avoid a whole series of problems. It was cheap, quick, simple, easy, and most importantly 100 percent effective for the potential failures identified.

A pre-accident investigation is exactly the same idea.

You and your organization can learn from this story. This is your guide to leveraging great, untapped knowledge that already exists in your organization. Your job is to prevent the frequency and severity of events in your company. I am convinced that the only way we can prevent events and failure is by learning. There is data to learn before an incident if you ask the right questions, and are willing to look.

Much like our scientist in the story, when things fail where I work we are extraordinarily smart, and can pinpoint exactly why the failure happened and what we should have done differently. Some of that knowledge surely comes in retrospect, but some of that information clearly existed before the failure happened.

There are many stories about knights slaying dragons. The successful knights were sure to investigate their dragon's weaknesses before they started their quest. The investigations gave each knight the knowledge to bring the right tools and skill sets to the dragon's lair. It was their investigations that laid all the groundwork for their success.

It is this type of information you want to discover in your organization.

Preface

> In a classic root cause analysis (RCA), our job is to deconstruct the event down to its most minute parts, analyze those parts, and fix whatever is broken. In Human Performance we do almost the opposite. Instead of deconstructing the event, we construct the event context and look not at the individual pieces of the event, but at the relationships between those pieces.
>
> From this book

You must have noticed over the last several years how our world is becoming more and more complex. The way we all interface on almost a minute by minute basis with technology in order to do our work is just one example. When was the last time you left the office without your cell phone or iPad? As our worlds and workplaces have become more and more complex, our failures of all types have also gotten more and more complex. Understanding how failures occur is no longer as straightforward as hunting for the broken part or the bad person.

In fact, we have now found ourselves in a world that is so complex it is no longer managed by our old, tried and true methods—methods that at one time were effective in reducing injuries, defects, and failures. This challenge is the reason why the new view, the Human Performance view of safety management, is becoming a vital next step in performance reliability management in many industries.

I am the Human Performance lead at Los Alamos National Laboratory. I have been a part of the introduction and slow, gradual change to the new view in an old culture. This has been a challenging journey for a facility as large and as technical as Los Alamos. My organization has been on this journey for some 10 years. We have made impressive strides in our safety performance. We have made even more impressive strides in the way we understand, talk to each other about, and move forward with learning from our failures. This is Human Performance.

Mostly, we have learned about our operations and systems. We have learned because we took the time to help our workers understand the importance of learning from each other, and in turn our management learning from the workers. I think we have much to share with other organizations about what has worked well, and what has been—let's just say less than stellar on our journey.

HERE IS WHAT WE ARE LEARNING

Changing your organization's safety management philosophy is a significant change. Change can be challenging, and sometimes hard. The type of change we are discussing in this book is especially difficult because you are asking your managers to demonstrate a new and different kind of trust in their workers. You are also asking workers to trust and communicate differently with their managers. This new philosophy will be hard at first. People will resist some of these ideas at first. But trust the process. Stay strong, and know that the work you are doing now will be some of the best work you will ever do in your career.

Hopefully this book will help guide you on this change journey. I have tried to give you some tools for successfully helping you get your message out to your organization. Where possible, I have tried to tell you in honest language what I have learned from introducing the Human Performance philosophy to many different types of organizations. I have always tried to sound as encouraging as I can in order to help remind you of how powerful this philosophy is to your organization.

Most of all, I wish you luck and success. Do extraordinary things. Change the world. Help build smarter, safer and kinder companies all over the world.

Todd Conklin

Acknowledgments

This book is the product of many smart people who do this work. Some of these people led the way for me (thanks for that), some of these people have been following me on this journey (you know I could be going in the wrong direction), and some of these people have walked beside me (nothing better than a good argument). All of them have made an important difference to the way I think about the new view of human performance. The interesting part of working with Human and Organizational Performance programs for the last 15 years is just how much of a journey it has been. All of us started someplace, often very different from each other, and interestingly enough found each other along the way. Big thanks to my fellow travelers: Shane, Bill, Pup, CKB, Sidney, Earl, Roger, Cindy, Kim, Rick, David, Doug, and everyone else.

Special thanks go to Kent Whipple for his strong editing sense and all-round smartness, and to my parents for their sense of direction.

I am anxious to see where all this goes.

1
A Story of a Failure

"Listen to the people doing the work...That's how managers learn..."

Management's Summary: A worker drilled into the top of a 5 gallon Liquid Propane tank with a ¾ inch drill causing a potentially fatal near miss of an explosion in a pressurized gas storage area.

Management's Action: Discipline the employee in question; send all shop personnel to additional training; rewrite shop procedures to include a section prohibiting drilling into pressurized 5 gallon Propane tanks with hand tools.

Freon comes in many different sized pressurized containers for many different reasons. Once used, any sized empty tank becomes a problem to remove from the facility (even before used, Freon brings with it its own set of environmental issues). The problem is that while the tanks are still under pressure, these tanks cannot easily, legally, or environmentally be removed. In fact, what is normally done is that the cans are punctured, flattened, and recycled.

For this story, the Freon tanks in question are 5 gallon containers. These Freon tanks are used all over the facility and, in a stroke of organizational and environmental brilliance, a central collection point was created in order to provide a place to gather all the empty Freon tanks when they had been used. This central empty Freon tank dumping spot was basically a square pin made of chain-link fencing with a gate. This area was specifically designated as a place for the facility to collect the empty Freon tanks, depressurize and flatten these tanks, and prepare them to be recycled.

This Freon tank collection system worked fairly well. No one person at the company owned the Freon process. The collection area and the process of recycling the Freon tanks was the "brainchild" of a team of workers who had been tasked with creating a way to recycle these tanks, minimize waste, and reduce the costs associated with disposing of the tanks. The maintenance organization kind of inherited this tank disposal program. It was not assigned to the pressurized gas department, nor was it directly assigned to the waste minimization teams. This program was given to the folks who managed the maintenance shops.

When the cage filled up with many empty tanks, someone was selected to go out to the storage area, sort through the tanks, depressurize and flatten the tanks, and pack them up on a pallet to go to the metal recycler. No one worker owned this job. This job was seen as a straightforward process and was usually assigned to new or less experienced workers. It was a straightforward task, requiring little experience, easy to do, and was seen as more of a nuisance than actual work.

For this event, a new worker who was not getting along well with coworkers within the shop was selected to clean out the Freon cage. He was taken to the work area and shown how to pop a small plastic seal on each Freon tank to be recycled, in order to depressurize and begin the process of flattening. He was shown the work activity by a supervisor, observed doing the work twice on his own, and left to finish the project alone in the work area. The worker seemed to understand the task, be able to accomplish the task with few problems, and capable of continuing with the task to completion.

The disposal of Freon tanks is an easy, unimportant, low risk task. This task was considered so unimportant that there was no process owner, no formal procedures, no identified risk assessment activity, and, therefore, no risk reduction activities created to do this work. Why would a worker need any of these things? Recycling Freon tanks is not a high risk activity. Well, that was the case at least until the process failed.

And fail it did. The worker drilled a hole in the side of a 5 gallon, pressurized Propane tank instead of a Freon tank. Propane is explosive. Freon is not explosive. Luckily, in this case the tank did not explode. In this case, the organization got lucky. This is a case of an event that did not happen, and in turn was a gift to the organization since they could learn from it without having a terrible failure.

However, the decision was made by the post-event investigation that this worker should have known the difference between the two types of tank. If the worker had recognized the difference between these two tanks, the worker would not have tried drilling a hole in the tank of explosive gas, placing his life at risk, and, as importantly, the future of the facility at risk. Nothing stops work at a company faster than blowing up a worker.

THE SECOND STORY OF THE SAME FAILURE

At this company's facility there was no disposal path for small, 5 gallon Propane tanks. You know these tanks pretty well. Every Propane grill in almost every back yard in the country has a small, white, 5 gallon Propane tank underneath. Many work groups at this facility owned Propane cookers. These grills were used for cookouts and picnics that were held in the same facility: worker celebrations, parties, and picnics. The Propane tanks on the cookers last quite a while.

Nothing is exceedingly unusual about Propane tanks and grills being used at a facility. The problem at this company started when the tanks became empty. In this company, it was immensely complicated to replace a Propane tank, or refill a Propane tank. A task that would take minutes to do if you were at your home was almost bureaucratically impossible to do at this facility. This created the problem of having empty Propane tanks at the facility, and no real way to refill these tanks. In fact, at this facility it was much easier to buy a new tank than to refill an old tank. That left empty Propane tanks with no disposal path available.

Unfortunately, the facility had no way to dispose of all of these tanks. There was no Propane tank recycling program. The pressure shop did not handle Propane tanks at any time for any reason. If you had an empty Propane tank in your work area, it was a problem. In fact, the pressure shop foreman said in an interview that an empty Propane tank at this facility was a problem for life. There was no process to dispose of these empty tanks.

Which led to an interesting part of this story of failure. It seems that because empty Freon tanks had a special little fenced-in area, an occasional Propane tank would not be seen as out of place there. After all, in the mind of a worker, a place to recycle a pressurized gas tank is a place to recycle a pressurized gas tank. In workers' minds all tanks are tanks when they have to find a place to deposit an empty one. So, after hours, when nobody was watching, employees at this facility were dropping off empty Propane tanks in with the Freon tanks. These tanks don't look alike, but they don't look that different either. The problem is that the tanks are entirely different. Freon tanks can be smashed and recycled. Propane tanks are almost impossible to smash and depressurize. They are actually designed to be refilled, not recycled.

To add to the complexity of this case, pushing in on a pressure-release seal button on the tank easily depressurizes a Freon tank. Propane tanks are not easily depressurized: even if you open the valve at the top of the tank all the way, it will not depressurize.

If your job is to clean all the white-painted tanks out of the storage area and depressurize, flatten, and recycle all of these tanks, you might find yourself in a position where in order to accomplish the boss's task you must adapt to any and all tanks that appear in the empty Freon storage area. More amazingly, the fact that this worker drilled a hole with a ¾ inch drill through the side of a Propane tank is quite remarkable. It is hard to drill a hole in a pressure vessel like a 5 gallon Propane tank. It takes time and discipline. It is not an easy task to accomplish.

THE STORY OF THE STORY

Suddenly, an event that was thought to have lots to do with a lack of experience, skill, and intellect on the part of this worker (how can that guy be so stupid? Geez nobody drills a hole in a Propane tank) was now a story about the procedure, ability, and policy for the disposal of 5 gallon Propane tanks at this company.

It is easy to say that the worker made an error in judgment in trying to recycle the wrong kind of tank, and, in fact, the worker did make an error in judgment of the situation; but the problem was not the worker's ability to know the difference between a Propane tank and a Freon tank—the problem is much more systemic, and more compelling.

The problem was a unique mixture of some pretty normal conditions at this facility. We can list many of them with little effort: a dual location of many small tanks in the same area, the inability to refill tanks at this facility, the inability to get rid of old and empty tanks, a new employee, the absence of 100% supervision, production pressures, performance management, HR—and the list can go on and on and on.

It is at this point that the facility will feel a need to make a decision. The facility will probably not know they are not making the decision they think they are making. The decision is not about how to handle the worker. The decision is far more about how the facility chooses to see this failure.

THERE ARE TWO CHOICES

Fix the worker (training, discipline, or termination) who did something he did not mean to do, in the hope that he won't again do something he did not want to do in the first place.

Fix the system that determines what should happen with the empty Propane tanks. A system that was clearly not designed to imagine the presence of Propane tanks where Freon tanks were to be stored.

Fixing the worker gives the impression of an immediate solution to the problem, but probably fixes the wrong thing. Punishing the worker is a fast and easy way to "solve" the problem, with the only issue being that it unquestionably fixes nothing at all, not even the worker in question. Because the whole failure will inevitably happen again with a different worker. It may take some time to drift to this point again, but over time it will happen.

Fixing the system that determines where empty tanks are stored and disposed of looks at a much larger problem that will take longer to fix, cost more money most assuredly, and take longer to complete. However, it will fix the right problem, and will ensure that the facility will never be in a position again to accidentally drill a hole through a Propane tank.

The second answer will be better for the worker, better for the organization, and better for the world, and will ensure a safer, more reliable path forward for this organization.

Sadly, it is *not* the path that most facilities normally choose to follow.

The goal for this book is to provide a usable road map for understanding a different way to manage safety in your workplace. In a way, this book should be more like a road map than a textbook. This book should answer one question: "All these ideas are great, but what should I do differently in my organization to make this change? How can I find, and solve these problems, before they become accidents?"

It is essential to note early in this book that I am standing on the shoulders of giants in this field. So much effort and thought have gone ahead of all of us in just understanding the theoretical underpinnings and creating a "new view" for the world of Human Performance and reliability. Many, many times I will direct you through quotes, cartoons, models, and ideas that are directly attributed to these fathers and mothers of the Human Performance idea. We won't spend much time on theory (although it is my hope that this book will make you want to read everything ever written about the topic), that is not the goal.

This book will give you practical guides, tips and tools, and suggestions for managing risks, hazards, and safety.

Practical means that the ideas and techniques you read in this book can be used in your organization almost the day you read them. It is the goal of this book to give simple and fast options you can deploy on the "plant floor" immediately. If you try some of these ideas, you will see changes immediately—seriously, immediately.

This is a substantial claim, a claim that in most cases I would be afraid to make to you. Not here, trust me.

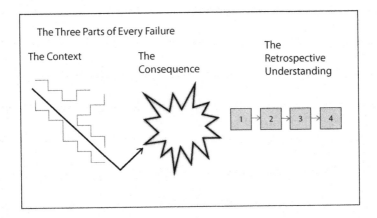

Figure 1.1   The Three Parts of Every Failure

WHAT IS THE DIFFERENCE BETWEEN THE WORD "EVENT" AND THE WORD "FAILURE"?

In our business we talk a lot about failures. Not only do we talk a lot about failures, we talk even more about events. In many ways, these words are used to mean the same thing—they are semi-interchangeable. In many other ways, these two words are worlds apart. We will need to give these two terms an understood meaning for our discussion, but first let's define what we mean by the word "performance," as in the title "Human Performance." Because to understand what is happening, a failure or an event, we must first understand how the idea of *performance expectation* sets the goal. Operational outcomes affect the meanings and context of a word. Performance, by its precise nature, is a word that is defined by its outcome.

WHAT IS PERFORMANCE?

> Performance is the degree to which you get what you expect from a person, a machine, or a process.

Anything other than what you expect is some type of deviation from what you wanted. Sometimes systems and people over-perform (although that seems rare, sadly), and sometimes our systems, and people underperform (that one you probably know intimately). Either way, performance is defined by the degree to which an action or process meets a desired outcome.

In reality, I don't like the terms "event," "accident," or even "failure" all that much. The best way to think about these "distinct moments in time or a process" that have either attracted or should have attracted organizational attention is as a moment where performance expectations were not met. Instead of using "event" or "failure" as words to describe these times, I actually prefer to use the term "deviation from an expected outcome."

Any performance expectation that is not met (over-performance or underperformance) is a deviation from an expected outcome. An accident, an event, even a near miss, is certainly a deviation from an expected outcome. A process improvement, a good idea, a success against all odds is also a deviation from an expected outcome. One is a bad thing, the other is a good thing, but they both are something different from what was expected.

DEVIATION FROM EXPECTED OUTCOME (DFEO)

The only problem with using the term "deviation from expected outcome" is that it gets really old, really fast. I tried to make an acronym out of this phrase

(DFEO); I never could get that to stick. DFEO just never caught on. I am not surprised it did not stick. It is awkward to say, and difficult to remember. It also doesn't sound very practical. Saying "deviation from an expected outcome" is something that a college professor would say, not a safety professional or even a manager.

So what is the difference between the term "event," and the term "failure"? The quick answer is that one of these terms has a big and bad outcome. The other has a small and simple outcome. But honestly the answer is that there is no difference at all in the words themselves. The difference in the meanings of these terms is in the person using the terms, and in the people that are listening to the terms. There is no difference in the terms, themselves—they are both deviations from expected outcomes. Both of these terms are indicators of a difference in expected performance.

A NEW DEFINITION OF SAFETY

For all too long now our organizations have tried to define successful safety programs by counting the number of times our workers screwed up or got injured. In actuality, what we count is the absence of accidents. Our safety incentives are measured and built around a zero accident goal. The Institutional safety numbers we report are usually numbers of injuries or days away from the workplace. These numbers are usually reported "up and out" of the organization, to customers, regulators, and the world.

You don't have to be a genius to know that something seems oddly wrong about the way we measure safety success. We count the number of people we hurt, and totally discount all the people we are keeping safe. The problem is, and always has been, you can't count what doesn't happen. It is hard to count the millions of decisions that are made every day in the field that don't lead to some type of failure. Those millions of operational decisions were all safety and performance reliability successes.

This new definition of safety or performance reliability demands not only a different way to think about events, accidents, and failures, but also, more importantly, a new way to manage your organization. You must change the way you manage safety in order to align your organization and operations to a new definition of what "safe" is, and why it matters. Assuming your management knows how to lead and manage this new method is a gigantic mistake, and will lead to horrible outcomes. These are new ideas and new ways of doing business. How would your managers know this information if you did not inform them of this new way of doing business?

Management now wants the safest workplace that is humanly possible. Obviously, no manager wants workers to get hurt. Knowing that management's intent is true, good, and hopeful, allows us to search for reasons for not changing other than motivation.

Those reasons almost always come back to one single factor. Most managers want to manage safety and performance using the best and latest ideas for success. The problem is that we, as a new-view safety professional community, have not told them about those ideas, and how to use this new view. Take the way we define safety—this is a perfect example of the old world view dominating our new world organizations.

SAFETY DEFINED

Safety is not the absence of events; safety is the presence of defenses.

Safety can't be the absence of events. Safety is the presence of defenses within the system, environment, and processes. If you constantly choose to use the traditional standards in order to measure safety success, you will constantly create workplaces where information about events and injuries is hidden from management. Worse yet, we are measuring our failures.

The problem is that it is hard to measure what doesn't happen. For every accident that occurs at a workplace, how many accidents didn't happen? How many people didn't get hurt? How many potential problems did workers who identified the error precursors before they had a chance to become full-fledged events solve? These numbers are seemingly impossible to count. That could be why these numbers are not valued by your organization.

In reality, safety is probably best defined by this idea:

Safety is the ability to perform work in a varying and unpredictable workplace environment.

This definition, although probably the most accurate, is the most difficult to use operationally and discuss with the workforce. Yet, we do this every day. Every day we live in an unpredictable and constantly varying world.

HUMAN ERROR

It is distressing enough that the world is varying and unpredictable. Human beings are also prone to their own unpredictability. Humans make errors. People are fallible, and even the best of us make mistakes.

Error is an unexpected deviation from an expected outcome. Error is an unintentional event. Error is doing something that you didn't intend to do. Errors happen all the time. Everybody makes errors, everybody. The very worst performers make errors. The very best workers make errors. Error is a predictable and natural part of being a human being.

People make a lot of errors. Not all errors have a consequence. In fact, not all errors are actually errors. We only really notice an error if it has some type of outcome or consequence that is large enough to be noticed by either you or other people around you. Error only becomes apparent if you notice an error.

Contemporary wisdom says that the average skilled worker, workers who work with their hands, makes 5 to 7 errors per hour. That same wisdom says that a knowledge worker, workers who work with ideas and concepts, not making things, makes between 15 and 20 errors per hour. True or untrue, we all make errors. Errors are how we are wired, how we are made, a natural part of being human. Human error is inevitable—all workers are error-making machines. What all this means is pretty simple: error is everywhere, and there is nothing you can do to avoid the errors. You can't punish error away. You can't reward error away. Error is an unintentional, unpredictable event. You know it, and I know it. Now we have to make sure that our organizational managers remember that fact.

But it is not that easy. Error is always attributed in retrospect to the worker by the organization after some type of consequence happens to the organization. In other words, the worker making an error is determined after an event happens, not before, and is usually seen as a moment in time where the worker did something wrong. If the worker had done something other than what the worker did, the event would not have happened.

This is always going to be true, but wrong, of every failure that will ever happen in your organization. The notion that the worker chose to make this error is also always true, but wrong. Choosing to make an error only becomes an actual choice after the event has happened. In fact, error can never be a choice. Error can never be a violation. Error is simply the unintentional deviation from an expected behavior.

It is easy to find errors in retrospect. It is even easier to judge these errors as wrong in retrospect. This process is called a fundamental attribution error. Think of it like this:

> A worker is walking across the office parking lot to go in to his office. During this walk, the worker steps on a rock and sprains his ankle. The injury is bad enough that the worker has to go to the clinic and have his leg treated. The worker ends up with a series of x-rays, a cast on his foot, and a day or two off from work.

This is an example of an error; however, this error will be attributed to the worker's judgment and walking ability. At some point, some manager will comment that if this worker had "watched where he was going" this event would not have happened. If this worker had cared more, the worker would have been more attentive. If the worker had been more attentive, the worker would have stepped over the rock, and completely avoided this injury. This injury is clearly the worker's fault.

The organization is attributing the error to the worker's judgment and behavioral choices. In a way, what the organization will do is assume that a bad outcome must happen to a bad person. You will hear some manager say something like this, "if only...if only the worker would have paid more attention..."

Read on and see how remarkably unfair that way of thinking is to your workforce.

> The next day, the manager of the "twisted ankle worker" who was injured the day before was walking across the same parking lot. The manager had a lot on her mind. She had a reportable injury. She had a worker who had to go to the hospital and get medical attention. She had an employee that had gotten hurt under her watchful eye. While this manager was thinking of all these things she stepped on a rock, the very same rock, in the parking lot, and twisted her ankle. She was in pain and could barely walk...but she didn't report the injury.

> Instead, this manager found out whose job it was in her organization to sweep the parking lots and sidewalks. The manager immediately called the roads and grounds crew supervisor, and had his crew sweep the parking lot. Because this manager was so busy thinking about the event that had happened to her worker, she was concentrating on her safety problem and not on where she was walking.

What is intriguing about this concept is the idea that when the worker walked incorrectly it was the worker's fault. When the manager walked badly, her response was to not only fix the problem, but also to move the fault from her as the manager to the person whose job it was to keep rocks off the parking lot. That is the fundamental attribution error, and it happens all the time with safety events. Every critique or after action meeting you will ever attend will have a moment when some manager will claim that had he been the worker he would not have made the same errors that were made by the worker in question.

Human error is a weird thing. Error is always present, but not always noticeable or noteworthy. Error is usually attributed after the fact—a way to explain rationally what happened and why. Error is hard to explain and hard to predict—these are accidental actions done without knowledge of the consequences. Error is never intentional—there is no such thing as an honest mistake because there is no such thing as a dishonest mistake—they are all errors, unintentional deviations from expected behaviors.

If all this is not complicated enough, the real question is if an error is actually ever an error. Error is essential to our discussion, as you will see as you read on in these chapters. Error, however, is not everything—I would caution you not to fall prey to the idea that if you just could stop error you would stop failure. You can never stop error, because errors we identify become visible only after the failure.

IS ZERO ACCIDENTS THE RIGHT GOAL?

The quick answer is you don't want to work in an organization that has any other goal than zero. Zero is probably the best individual goal that any worker could have. What is more compelling is the fact that most workers actually meet that goal. Most workers will work their entire careers and not have a serious accident or event. Zero accidents actually becomes in reality an outcome, the right outcome, but having zero accidents does not make a safety program. It is, at best, the target at which your program is aiming. Asking workers to have zero accidents does not tell workers how to have zero accidents.

The problem is that "zero" is not a particularly realistic goal for the entire organization. In short, the problem is that "zero" is a standard of perfection. Your workers and managers are not perfect, and will make mistakes. They know this fact, and you know it. You will have failures. You will not be able to hold events to a "zero standard" over the lifetime of the organization. You will have failures.

In many ways, that's why your organization has learned to count the number of failures, and track that number—with the goal of constantly driving that number down to zero. It is a noble and noteworthy goal, zero. The only problem with having zero for a goal is that you (and every other safety professional in the world) are only one ankle sprain away from being the worst safety manager at all times.

There is another problem, and that is the problem with using "perfection" as our goal. Perfection leaves no room for the inevitable human error that will happen. And we know it—errors will take place. We just need to learn where and when they could take place. When your organization uses only perfection as its standard, anything other than perfection is just fundamentally wrong. Let's not belabor this point too much; you live this every day of your life, but suffice to say that if a victory in your business is no accident, we will never actually know how you won the contest.

Humans and organizations learn by trial and error. We learn by becoming smarter after we fail. In fact, when we fail, we most often go right back out to the worksite and try again—only this time we try to fail smarter. When we devalue failure, in many ways we are inadvertently devaluing our ability to learn.

WE CAN LEARN A LOT FROM THE AUTO INDUSTRY

Almost every day in some workshop someplace, I ask this question: "How many people were killed in fatal highway accidents last year?" It is a depressing question to ask a group of workers. Every highway death is a tragedy. But this is a compelling question.

Why is this compelling? Partially because in the United States the fatal highway numbers have stayed right around 38,000 people a year for almost 50 years. In 2000, our numbers were around 38,000. In 1990, the numbers were around 38,000. In 1960, the numbers were also nearly 38,000. We have leveled out in the number of people killed on our highways.

Yet, the number of cars on the road has dramatically increased. The number of drivers has increased. Also, the number of millions of miles driven by American drivers increases every year. Still our highway fatality numbers stay essentially the same.

Why?

Every day, workers tell me the answer is that cars have changed. Cars are safer today than they ever have been in automobile history. Today, roads are safer, better designed, better engineered, better marked, better patrolled than in the past. Everything in our highway and vehicle system is better engineered, smarter, and safer. In fact, the auto industry has done everything but change the driver.

The Statement "change everything but the driver..." is important. In many ways on our job sites, we have taken the opposite approach. We have tried to get to safety performance by "leaving everything the same except fixing the worker." It is about time to admit to ourselves that we have been managing safety wrong. We were simply doing safety management backwards.

We can learn much from the idea that our organizations respond in the opposite way. We have almost always "changed nothing but the worker." We rarely fix the system around the worker. We almost always try to "fix" the worker—as our sole corrective action.

Why is it safety programs in our organizations are so interested in fixing the worker and not truly engaged in creating survivable space in our work systems and processes? I fear that the answer may be either we are lazy, or we are cheap. All defenses in vehicle transportation that can be engineered to create the least number of failure consequences possible—or the most survivable space possible—are being created every day with one goal: "We cannot fix drivers, drivers will never be perfect, drivers will screw up. We must design and create systems around the drivers and passengers that keep them safer."

Drivers, according to the daily discussions I have with workers and managers in classes, and workshops, are getting worse. There are many things that make drivers less attentive, less effective, less safe in their operation of a vehicle. Cell phones, texting, GPS units, fancy stereos, iPods—the list can go on and on of things that steal attention away from the highway. In fact, the automotive industry assumes that drivers will be placed in positions of trading attention between driving and "functioning" in the driver's seat.

This is most compelling, because highway fatalities are getting rarer and rarer. In fact, chances are extremely high that you will not die while driving or riding in a vehicle. This is a direct result not of trying to stop car accidents (which are all bad outcomes and should be avoided), but, in fact, of knowing that when an accident inevitably happens the goal is to reduce the consequences of the accident. Make cars fail safer.

Making our cars fail safer? That is exactly the primary idea of defending against accident consequence. Not one person has the power to stop all accidents in the workplace. Human beings simply are not particularly good at predicting the future. I will gladly admit that I am not smart enough to avoid all accidents in my personal life, let alone in my facility.

Because you cannot prevent all accidents, you must assume that accidents will happen, and use your time, energy, effort, and resources in dramatically reducing the consequences of the accidents that will happen in your workplace. You must build systems that allow our workers to fail safer. Start thinking like the car industry, and make your systems safer.

RUMBLE STRIPS

Probably my personal favorite defense against vehicle accident consequence is the humble rumble strip (pardon my rhyme). Rumble strips are the grooves that are placed in the actual road surface along the roadside, just outside the normal driving system, that alert the driver that they have drifted out of the system. These strips vibrate and audibly alert driver that something in the system is different.

Rumble strips assume that the driver will make a mistake. Rumble strips are designed to work when the driver screws up. In fact, rumble strips will *not* work when the driver is actually behaving as road designers assume drivers should behave. Rumble strips only work when the driver becomes inattentive.

What rumble strips do is allow the driver to have a reasonable warning that something is wrong. The driver may then correct their behavior before an accident happens. The screw up is still there, the only difference is that the driver has time to recover. This example illustrates the importance of this idea.

Workers (and drivers) are at their best when they are allowed to detect and correct their performance in normal systems. Normal systems are most reliable (and they are by definition terribly reliable) when the humans using those systems are allowed to detect trouble, and then correct that same trouble while in normal operation of work.

FROM TODAY ON...LOOK FORWARD

It is not that your managers, workers, and safety professionals are bad, or that they make bad decisions. That is simply not true. Your managers have done and continue to do their very best in making the workplace safer and more reliable. Managers assuredly do want what is best for the organization, and the fixes and corrective actions that have been done for years were fixes and corrective actions that managers believed would fix the problem and make the workplace safe.

In fact, it is essential to remember that managers have the best intentions in mind when they march workers off the job site, or when they fire a worker for making a mistake, or when they scream at workers for not stopping work right before the failure that just happened.

The problem that we are attempting to solve with this new Human Performance philosophy is not in how we are motivating owners and managers; the problem is much more about teaching a new way to manage safety in your organization. Just as we say it is bad to try to "fix" your workers, it is just as bad to try to "fix" your managers.

Consider this: you are not trying to change the managers in your organization, you are just trying to move their understanding of safety and reliability to a different perspective. These managers will still be able to use their talents, experience, and management tools—they will just be using them in a different way.

What you are actually trying to do is to move your organization from a "crime and punishment" model of managing safety performance to more of a "diagnose and treat" model of safety management. We know that we can't punish away safety issues. Try as we might, we cannot hold workers so accountable that they never again have a failure. We have to identify and understand the underlying issues that create the environment in which these failures can happen.

Don't go back and look at old investigations and corrective actions to see how you could have responded better or differently. That only serves to bring up unpleasant feelings around events that you worked hard to both understand and "fix." It is far better simply to draw an imaginary line in your calendar today, and simply respond differently the next time you have to respond. Look forward, not backwards, in managing safety. You will have plenty of chances to try these ideas. Just as you will have many chances to talk your managers out of old school reactions to new view responses. In our business we always have opportunities to practice these skills.

Try these ideas, and ask your management team to try these ideas. You honestly have nothing to lose and everything to gain. Remember, you can always move back to "blame and fix."

2
Why Think about Failure at All, Let Alone Think "Differently" about Failure?

QUESTIONS ABOUT YOUR ORGANIZATION:

- Is your organization as safe as it should be?
- How does your organization measure safety success?
- Is the production of product more valued than protection of people, really?
- Look around and see what is valued with money, time, and recognition— what is most relevant to your management team?
- Are your processes written to make work happen successfully, or to avoid compliance failure?
- Do you learn as an organization from your successes as much as, or more than, from your failures?

Here's what we have learned about safety...not everything, but some crucial things that seem to make a difference at work. If you will allow me the pleasure, I am going to make a list and, in turn throughout this book, discuss these ideas in more detail:

- The safest workers seem to be workers actually performing work.
- Workers are as safe as they need to be, without being overly safe, in order to get work done.
- Workers consistently create safety in practice while they do their work.
- Workers are constantly detecting and correcting variations in the work, the work environment, themselves, and others—to create safety.
- The work the workers do in an organization is remarkably different from the work that was planned for the workers to follow.
- Workers constantly have to make the work match the process, not the processes match the work. Workers are not empowered to "officially" change your process—only you have the power to change the process.
- Your worksites are undoubtedly a "hidden laboratory of alternative choices," in which we are always hoping that the workers make the right choice—and normally they do...until they don't.

- Your job sites, work areas, and facilities are normally exceedingly safe and dependable—worse things *can* happen than *will* happen—and because work is normally safe, workers are less careful.

In the opening case study, the Propane versus Freon tank failure, the failure is so damn stupid that it is hard to believe there would be any other organizational reaction to this failure than terminating the worker. The worker in this case study had it all—bad attitude, lack of ability to get along with his coworkers, "stupid" written on his forehead; he probably had bad habits and smelled funny. And in many ways this is the precise problem we are trying our hardest to combat. It is faster, cheaper, and a lot easier simply to claim the problem belongs to the worker and "fix" the worker. By simply choosing to purchase this book, you are telling me that you already have thought about why "fixing" people is so remarkably risky.

When you think of Human Performance, think of the three layers of your organization that are always present, during both failure and success. All three layers must be considered:

- The individual worker
- The organizational system
- The performance expectations set up by management

Too often organizations tend only to consider the first third of this list: the worker. The worker is where we put most of the effort; the worker is what we try to fix. The problem is if you only try to fix the worker, you are only getting ⅓ of the total organizational landscape. You therefore miss ⅔ of the potential problem: the system and the management expectations. You must teach yourselves to look at all three levels.

If the organization in the case study had fired the worker, it is a decent bet that in this case the problem would have gone away. The idea that there could ever be another perfect combination of a storage area, a tank smasher, that worker, that supervisor, that day, and that particular organization is incredibly low. But it is still there.

Is a "chance that event will happen is extremely low" an operational limit you can live with in your organization? Could your organization survive two failures involving Freon tanks being smashed dangerously?

By understanding the entire story, the whole system that "housed" the event, you can begin to understand not only the failure in the case study. You can also understand the other failures that are waiting to happen within that system, that organization, even that same work team. The ability to investigate an accident before it happens seems to be an extremely positive step forward in the understanding of not just failure, but the systems, people, technology, and components that must be present for failure to happen.

THERE ARE LOTS OF WAYS TO BE SAFE, BUT FAILURE SEEMS TO HAVE A SINGLE PATH

This part gets a little tricky. My advice is not to over-think these next ideas, but to understand how vital it is to your future success to recognize that these ideas have power over how you react and respond to failure. In short, the failures you look at in your organization could have only happened the way that they happened. If anything else had happened during the event, the failure would not have taken place. This idea is always true in retrospect, and seems to need the addition of the word "duh!" every time it is said in a meeting or event review.

But ponder this idea for a moment. Knowing the nature of all the things that had to have taken place in order for failure to happen is pretty valuable knowledge about not only the failure event, but also the organization that hosted the failure.

How much of what happened always happens?
Answer: Almost all of what happened always happens.

How much of what happened has never happened before?
Answer: Usually nothing different is recognized until after the accident happens. Remember this: it is exceedingly hard to notice something that doesn't happen.

How relevant is it to know the difference between the two sides of the Freon/Propane story: the first story and the second story?
Answer: Knowing both sides of the story is the most vital part of understanding the event. In many ways our entire job is to be able to look at both sides of normal and abnormal factors, and make sense of this information for the organization.

We are often lulled into believing that if we can just find the single, root place where the world went wrong we will have done our job and will then be able to make work a safer place for all. The problem with this idea is that we are learning that there is no one, single root cause that is the problem. The problem is always the relationships, causalities, or spaces around that single root cause, which seem to be in a constant state of motion and variability.

Knowing what was normal, what was not normal, and the entire gradient between normal and not normal begins to tell a different story of the accident, incident, failure, error, mistake, or upset.

WHAT IS FAILURE?

It is pretty essential to get our vocabulary together. So let's start with the understanding that when we use the word "failure" we are actually meaning the universe of words we can use to describe a deviation from expected outcome—DFEO. Failure is a good word to use because it is a large word that encompasses many other ideas. Accident, event, mistake, operational upset, flaw, deviation from the norm, screw up, foul up—these are all words that fit within the notion of failure. So when you see or say the word failure you are opening up all of the above mentioned situations in a quick and efficient manner.

Failure as a term is also significant because it crosses professional boundaries well. Quality professionals understand failure, production professionals understand failure, budget professionals understand failure, and, therefore, you don't have to spend much time translating these ideas internally.

Perhaps as significant, failure as a word translates well between professions and between cultures. There actually is not a direct translation for the word "safety" in Spanish, Japanese, or many Middle Eastern languages (all cultures I have spent much time working with Human Performance issues with in the last couple of years). But the word failure moves easily between concepts and languages in many cultures.

Why some cultures don't have a word for "safety" and do have a word for "failure" is, to a great extent, a real fundamental part of this book, and is pretty darn interesting. Safety, in many ways, and in many parts of the world, is a luxury that exists when the rest of societal needs are met, or close to being met. It hardly matters if the workplace is safe if you are more prone to being killed at home or on the streets. The luxury of safety for those of us in the developed world is no luxury at all—it has become a necessity and now a true business advantage.

Let's further define failure using Eric Hollnagel's definition. Hollnagel, an expert in this field, says, "A failure is the unexpected combination of normal performance variability." Hollnagel goes on to discuss that accidents don't happen because workers in the field have gambled their lives, and the organization's reputation; failure happens because of three key ideas.

First, failure happens because the worker believes that what is about to happen to them is simply not possible. This is so profoundly true that it seems simple and obvious. However, after a failure takes place we so easily move to a place where we see the entire picture. We know how the story started, what happened in the middle, and how the story ends, so we can see every possibility for the failure along the way. The worker has none of that information, and, therefore, does not have the same story or view of the failure within the event that we hold after the event.

Every time you drive home from work you fall into this potential failure mode. I will bet you that you don't perform a major safety systems check, take a pre-drive walk around your car, or even check your mirrors before you go home. I will bet you hop in the car, and drive home. That is what most of us do every single night of the work year. Why? Don't you know better?

Sure, you know better, but you also know that that same car got you to work in the morning without failure, so why would it not get you home that night? You have no indicator that anything is wrong, or even could be wrong. So you just hop in the car and go. Workers do the same thing at work. It worked last time. It worked the last 10,000 times. Normally, it always works OK. Why would it not work the next time?

All of us know that success one time does not lead to success the next time. Yet, success one time actually does usually lead to success the next time. In fact, failure is extremely rare. Your car does not break down very often. In fact, your car is much more dependable than it is not dependable. Think about this idea for a moment. Your car runs many, many more times than it doesn't run. Your car is more successful than the occasional mechanical failure. The same goes for workers.

Secondly, failure happens because failure often has nothing to do with the tasks and processes that the worker is currently doing in the field. We focus exceptionally well as humans on one thing at a time. Current neuroscience is telling us that humans are not particularly proficient at multitasking. In fact, neuroscience studies show that humans don't multitask—instead, humans share attention, some of us are fast attention sharers, and some of us are slow attention sharers. Depending on the context, we often find ourselves in situations where we cannot share attention at all. Think of the moment you pull a rental car out of some foreign airport and have no clue which way to go. At that moment, you are focused entirely on one thing—driving in a direction that will get you to your destination. You stop attention sharing, you turn down the radio, and you are not talking on the phone. You are driving for your life, or so it seems.

That same notion is true of things like texting and driving, or welding and catching your clothes on fire. These things are connected, but they are usually connected post-failure and in such a way that the connection becomes obvious and simple—after you see the connection.

Thirdly, failure happens when the worker feels the possibility of getting the intended outcome is well worth whatever risk is present in the work environment. The worker makes a trade-off between being safe and being productive. You want productive workers. The worker chooses to take a chance and guess that this iteration of this process will not fail this time and in this way. In other words, the worker looks around to see if any manager is watching and goes for it. The pop culture reference to this is "get 'er done!"

Many of you are fully prepared to argue that that is gambling. It may appear to have all the components of gambling. It may look like gambling. Hell, it probably smells like a gamble. Problem is, it's not gambling at all.

When you think of worker intent, or, as we have discussed it so far, gambling workers, think of how serious the outcome is to the notion of risk. In fact, think of this by using one simple word: surprise!

If a worker looks around and takes a chance and it fails, and the worker is *not* surprised, then the worker knew better than to take the chance, and probably was really gambling with your organization's reputation. But, if the worker looks around takes a chance and it fails, and the worker is genuinely surprised by the failure, then the worker was trying their best to get a difficult task done. It honestly is that simple, and is the greatest test of intent that I know personally. What do I normally discover when I am sent in to investigations? The surprises. I find many places in the middle of failure contexts that are just plainly and simply a surprise to the worker. I use the word surprise in my write-ups of events. I find the word and the idea of a surprise condition that the worker discovers while doing work to be a concept that is well understood by the people doing the work, and also well understood by the people who manage the people who do the work.

TYPES OF FAILURE

There are primarily two types of failure. These two types of failure are noteworthy because they help us understand how to move forward with the process of understanding not only how to investigate a failure, but also how to create corrective actions that will ensure that this failure, and others like it, is less prone to take place in our organizations. Knowing the type of failure allows us the power to know how to proceed in organizational understanding and learning. Soon we will discuss the power of learning in creating safe and efficient workplace systems.

- **Individual Failure**: A failure happens and the worker is not protected from the dangers present in the work environment. The consequence of the failure is almost always to the worker or workers in the event. This category of failure would include events similar to cuts, slips, trips, and falls; chemical or hazard exposure; strains. In general, failing to protect the worker from anything physical happening to that worker is an individual failure. Hurting people is an awful thing. Hurting a worker is seen as a failure of the worker not to operate safely.
- **Organizational Failure/System Failure**: A failure where the organizational systems allow some type of threat to the system to have a consequence where many people are adversely affected. In a system

failure, someone or something has been able to break through the many layers of defenses that were thought to be in place protecting the facility, its people, and reputation. What you often find is the profound difference between thinking there are many layers of defenses, and actually having the many layers of defenses in place. Organizational events are things like oil spills, reactor coolant failure, single point recycling storage pens for collocating different small gas containers, or even air traffic controllers sleeping during the nighttime hours when no planes are scheduled to land at an airport.

Why do you care about these two failure types—and I might add you certainly do care about these two failure types? Because failure is almost always much more complex than it may at first appear.

Dekker says it something like this: it is much more seductive to see Human Performance as puzzling and perplexing, as opposed to complex. We become much more interested in the "why" question and somehow miss the "how" question about failure. Where we see puzzling "whys," the story of the failure often tells us something quite different. The story of the failure finds that success is complex: safety critical work depends on expert human performers performing work in an expert way. Our organizational systems tend to run degraded (from the launch of a system until it runs to failure), and plans, processes, rules, regulations, and procedures that we assume guide work towards safe operations are almost always incomplete.

You use the types of failure—something you have done automatically your whole career to understand where to start and stop the telling of the story of the failure. Most of the stories I tell in my work are stories about the way the organization failed the worker, and almost never about how the worker failed the organization.

The concept gets clearer when you break a failure down a bit.

> Caution: I am not asking you to deconstruct the failure down to its smallest parts. Near as I can tell, examining an event by taking the event down to its smallest parts has little value at all for you, for the workers involved, or for the organization. What we want to deconstruct is not the failure, but the way you and your organization understand the failure.

This next idea will help a lot.

FAILURE HAS THREE PARTS

Every failure that takes place in your workplace can be divided into three distinct parts. Let me rephrase that statement: failure must be divided into three parts in order for you and your organization to begin to understand what is happening. You must be able to understand how the failure environment changed while the actual failure was happening.

The three parts of a failure are straightforward, and immensely important:

Part 1. The Context: Everything that led up to the actual failure event

We will call this "the context of the failure." Sidney Dekker, my friend and challenger on all things involving my organization, calls this part of a failure "getting into the tunnel with the worker." Everything—and I do mean everything—that is happening while the work is being done plays a role in the outcome of that specific work. We must try to capture and understand all the things that are happening to set up this failure to fail. The problem is that we must look both at the timeline of the event (a rather linear view of the world and not at all truly representative of reality), while at the same time look at the complex relationships that exist between all the many moving parts that had to align to cause this failure to actually be triggered by these workers.

Part 2. The Consequence: The failure itself

Someone or something triggers an unexpected outcome that for our discussion is seen by the organization as a bad outcome. Oddly enough, this is both the easiest part to identify and the most uninteresting part of the failure. This actual failure shows itself to us in many ways. All of these ways will be a clear deviation from an expected outcome. Often these ways are tragic and horrible. But the failure itself is only significant in the sense that it is the event. The failure will have an immediate consequence and will drive an often immediate organizational reaction.

Part 3. The Retrospective Understanding: Everything that happens after the failure happens

Call this the organizational reaction. Ah, the power of retrospective. This is where understanding the three distinct and different parts of a failure presents its best and most valuable pay-off. Everything that happens after the failure is colored with one giant rational advantage—that is, the fact that from this point on you and your organization know how this story ended. Remember, those workers back up in Part 1 of the failure had no inkling of this knowledge. Had they the rational advantage of knowing this failure would happen, it is a pretty solid bet that they would have done something—anything—differently in order to avoid the Part 2 of this event.

All three of these parts of a failure *always* exist in every failure, no matter the size. All three must be recognized and understood. Perhaps most importantly, you and your organization must be able to understand that all three of these failure parts are different, and must be defined and understood as different in order to understand fully the "story" of the failure itself.

WHEN UNDERSTANDING FAILURE, SOME THINGS ALWAYS MATTER

All divisions of failure aside, there are a set of core factors that seem to be present and relevant in all events. These core factors are necessary and collectively sufficient to help you tell the story of the failure. Without the core context information, your story is incomplete.

Core Context Information List:

I must have an explanation of the failure
I don't need the causal analysis document; I don't want the fault tree; I am not even sure I need a timeline; but I have to have an explanation of what happened. I am surprised by how often I go to the field and look at amazing report documents that don't thoroughly explain how the event happened. I actually look for a story that can be told of how the failure happened. Remember your junior high school English class? A story has a beginning, middle, and an ending; the story must make sense (have fidelity), and move through time logically and sequentially, but not necessarily linearly.

I need to understand how the energy of the failure was or was not isolated from the worker or workers involved in the failure
That sentence is written in High Reliability-ese; let's see if we can reframe this idea in a way that makes sense to people who are in the field. I need to know what went wrong and what went right. Where did the failure start, and when did our systems kick in to mitigate further failure (if at all)? Which all leads to the third requirement...

I need an understanding of how some type of defense or defenses built into the organizational system were either not present, or if present for some reason failed
This chapter of this book seems to contain many absolutes. Here is one more: every failure that takes place happens because some defense either did not work or was not present in the work environment. This defense comment is always true; albeit incredibly retrospective, it is always true. Think of it like this: I am not surprised that air traffic controllers, working alone in the middle of the night at an airport with no scheduled landings or takeoffs, take naps. I

would be surprised if there were no way to wake the controllers up when they were needed to perform their mission. A horn or a buzzer connected to the radio would be an excellent defense against an exceptionally severe lack of stimulation and an incredibly high amount of boredom.

I must be ever vigilant in remembering, "size does not matter"
The consequence of a failure does not determine the importance of the failure. We are humans, our organizations are made up of humans, and we all believe that consequence drives outcomes. I have spent the last 15 years of my career asking this question of classrooms full of people just like us: "What is the difference between a big event and a small event?" The answer, and I think I have heard them all, really should always be, "nothing." All events allow us the opportunity to learn. Small events, those events with small consequences, are often much richer in context and story than what our organizations define as large events.

I must always remember that bad things don't always happen to bad people
Or at least stop thinking the person is bad (or allowing the person to tell me that they were for some reason bad). The truth of the matter is that bad things happen all the time to very good people, very experienced people who had no intention other than doing the very best job they could do for their organization. As silly as this seems, not thinking people involved in an event are bad is difficult and will take deliberate attempts on your part to not label the workers involved as something less than perfect.

Here is what you know from years of doing safety work. Sometimes big, horrific events are actually quite simple, straightforward failures. A bolt snaps, scaffoldings fall, and the consequences are terrible. The other side of that coin is that sometimes tiny events have profound learning implications for the organization. A small event happens, and you suddenly realize your entire system is flawed and must be immediately strengthened. Event size is not nearly as powerful as event learning. Don't let the drama of an event (or lack of drama of an event) dictate how much or how deeply you respond.

The problem is that the only way you know which events are information rich is to look at them. Therefore, every event demands event learning. Every event must demand event learning.

Case Study
The *Titanic*: A Story to Help You Rethink How You Think About Failure

Before you read this case study, a study you will undoubtedly know something about, ask these three questions as a pretest of your failure understanding ability:

- Whose fault was the sinking of the *Titanic*?
- What caused the *Titanic* to sink?
- What corrective actions would you have recommended for the next ocean liner to be built?

The *Titanic* was built to be the flagship for the White Star Line. It was the largest and most opulent passenger ship in the world. This stately ship was "state of the art" for its time. White Star designers touted a ship built with all of the luxuries of the "gilded age," with countless fineries and undeniable craftsmanship: a Turkish bath, teak wood furniture (even in third class), a heated pool and lending library. There were crystal chandeliers, hand carved banisters, and gold leaf detailed throughout the ship. No expense was spared in its luxury and beauty.

The same eye for detail was brought to the technological design of the *Titanic*. New technologies were integrated. It boasted a 5 kilowatt Marconi wireless radio (invented just 11 years before), a telephone system, two bronze 10′ propellers, and a double plated 2′ steel hull. The builders and designers had believed they had built the safest ship in the world. In a word, it was "unsinkable."

THE SAFEST VESSEL OF ITS TIME

The *Titanic* was to be the safest vessel of its time. A major improvement to its design was its hull. The hull was subdivided into watertight compartments. This was a new safety design and had never been done on any other passenger vessel.

The ship was monumental, measuring 175' high, with 9 decks. At its top speed it could travel 23 knots (26 mph). Fully loaded it had a passenger capacity of 3247 in all three classes, and a crew capacity of 885.

The *Titanic* began its maiden trip from Southampton, England to New York City on April 10, 1912 with 2,223 crew and passengers. The first class passenger list boasted the "jet set" of its time. The list included Benjamin Guggenheim, John Jacob Astor IV and his wife Madeleine, and Isidor and Ida Strauss, the owners of Macy's department store. In the second and third classes were many families moving to start a new life in America.

SYSTEMS ARE NORMALLY DEPENDABLE...

The first few days of the journey went incredibly smoothly, leading the ship's owner, Bruce Ismay, and its experienced Captain, Edward John Smith, to begin to feel a bit more at ease. They considered pushing the ship's speed a bit. The normally choppy North Atlantic seas were calm, and the weather was cooperative. Every hour the crew would drop a bucket over the side of the ship and bring it up, drop in a thermometer, and measure the temperature of the water. The temperature would indicate the presence of ice. The crew were rotating their shifts in the eagle's nest at the top of the ship, spying for ice. The ship was also receiving Marconi messages from other ships in the area providing the location of ice fields moving into the shipping lanes. These were the standards of practice at the time. The crew was making sure that they, and more importantly, the passengers, would have a smooth trip to New York.

UNTIL THEY AREN'T

In the late evening of April 14, 1912, right after the first class passengers enjoyed a magnificent 10 course meal, at 11:40 pm, the *Titanic* hit an iceberg. Only seconds after the crew saw the ice, the ship was hit on its starboard (right) side, with the impact lasting a few seconds, leaving a giant gash in the steel hull.

In just under three hours, this dazzling, beautiful liner sank to the bottom of the North Atlantic, resulting in the tragic demises of 1,500 people. This is the deadliest maritime disaster in our peacetime history.

HOW?

How did this happen? This ship was "unsinkable." There were systems in place to ensure everyone's safety. The *Titanic* was a technological marvel. It had everything to conquer the predictable hazards of transatlantic passage at the time. Multiple safety systems existed around every analyzed hazard that was known. Safety was not just a significant value; safety was also a marketing strategy. It would be hard to find evidence that reliable and safe operations were not one of the most important operational values. Nothing could have or should have ever happened. The entire mission of the *Titanic* was not to sink, and yet it did.

Let's continue the story of this complex, multilayered failure. All of the designed safety systems, and trained and safe workers were in place, and functioning exactly the way they should have been. Nothing unusual happened; in fact, what happened was an amazing and unexpected combination of many normal things.

We will never know the exact size of the iceberg; however, early newspaper reports of the time estimate that the iceberg was around 50 to 100' high and 200 to 400' long. However, an interesting point about the size of icebergs is that only a teeny, tiny part of the berg is visible above water; the greater part of the iceberg is underwater, and is 10 times larger than the visible portion. Most icebergs expose about 1/10 of their mass above water, leaving the other 9/10 of their mass below the water line. In this part of the Atlantic, icebergs were an expected, accepted, and normal part of sea travel.

Sixteen separate cells of *Titanic*'s hull could each be isolated and closed with special doors in the event of an emergency. This was an amazing new design for any passenger ship. However, to be watertight they would have to be able to be closed like a box on six sides, making parts of the ship impassable to crew and passengers. This unintended impediment was operationally impractical, so doors were placed in the bulkheads to allow access. Because of these doors, an adaption in the ship's design so that the ship could be operated, the watertight bulkheads were not really watertight. Because of the design confidence in these separate cells, no consideration was given to the idea of what would happen if water did penetrate them. The cell walls were tall but not watertight all the way up to the ceilings of the ship. Water could flow over into each hull compartment. All of these new and highly technical defenses were designed with the best intention of safety. This system was the most modern and the safest ever developed.

In the crow's nest of the ship, the place high above the ship where crew kept a lookout for ice, there were supposed to be a pair of binoculars. White Star leadership made a relatively normal last minute change in the equipment officer. The old equipment officer forgetfully took the key to the equipment storage locker with him when he was transferred from the ship. *Titanic*'s new equipment officer couldn't find a replacement key, and because there was no access to the equipment storage area, the lookouts were not issued with any binoculars.

The White Star Line's official policy was that binoculars were supposed to be given to all lookouts. Since they only had a few binoculars on the ship due to the error, these binoculars were given to the highest ranking officers, not the lookouts. On the night of the collision, the weather was cooperating. It is very difficult to notice something that doesn't happen, and in this case in this sea condition, nothing was happening. In these conditions it was very difficult to see the iceberg. Churning or violent water would have caused breakers, with some flotsam and jetsam, around the base of an iceberg in the water, making it easier to see it from a greater distance.

Titanic did not strike the "typical" iceberg. It wasn't white, gleaming out of the ocean like most icebergs. Witnesses say that it was almost invisible from continuous melting and refreezing. This clear surface of the berg reflected the water and dark, clear night sky like a mirror, thereby making it almost black, or the same color as the evening ocean. It would have been the ocean equivalent to the black ice seen on roads. This type of iceberg is nearly impossible for a qualified and competent lookout to see in these conditions, even with the keenest eye and the highest professional motivations. The very best lookout could and did easily miss this hazard. Not because the lookout could have done it better, but because this hazard was so difficult to identify in that condition.

The *Titanic* crew had a total of 6 hours for sea trials before they sailed from Southampton to New York. The sea trials were pretty much a formality. The members of the crew who stoked the fires were on board, and representatives of the White Star Line, but no domestic staff were aboard. Once the surveyor from the British Board of Trade, who was also present, saw that everything worked to his satisfaction, and that the ship was seaworthy and crew were fit to service passengers, he signed an "Agreement and Account of Voyages and Crew." This document was valid for 12 months, and that deemed the ship seaworthy. They did have additional time to prepare when the ship traveled from the shipyard in Belfast, Ireland to Southampton. The crew had 6 hours of practice to prepare for its first intercontinental cruise. Many of the crew did not know where stairwells and exits were located, especially in the second and third class areas. Some of the exits were locked, keeping the passengers below deck. Familiarity with the ship was clearly as complete as it needed to be without being overly complete. This short sea trial was perfectly acceptable for the most technically advanced and safest ship ever built. This sea trial only becomes incomplete, rushed, and wrong because the *Titanic* was lost at sea.

SAFETY SYSTEMS COMPLIANT

The *Titanic* was fitted with a total of 20 lifeboats, comprising three different types. Initial designs for the ship had included 48 lifeboats. However, in an uncharacteristic but financially reasonable cost-cutting measure they cut 28, leaving only enough lifeboats to hold 58 percent of the passengers and crew. Because of the added safety design features built into the ship itself, it was acceptable to determine that the additional lifeboats were not necessary. The White Star line was fully compliant, well within the law. The British Board of Trade regulations stated that all vessels over 10,000 tons must carry 16 lifeboats with a capacity of 5,500. These regulations were becoming antiquated since the sizes of passenger vessels were growing enormously. The *Titanic* weighed in at 46,000 tons. Not only had the White Star Line fulfilled their legal and regulatory commitments, they were providing more lifeboat accommodations than were required. The high casualty rate in the sinking was due to the fact that although White Star were meeting the current regulatory requirements, the Titanic only carried enough lifeboats for 1,178 out of a passenger and crew list of 2,223. An outcome from the investigations into the disaster in both England and the United States was a drastic change in the antiquated regulations.

Recent tests on the steel from the *Titanic* reveal that it was far more brittle than the steel used on modern ships, but it was the best available at the time. It is easy to look back and say that they used inferior steel. However, that is the advantage of time, retrospect, and knowing the ending of this tragic story. The White Star Line spared no expense in this part of the construction, and used what was the best material of the time.

PROFIT OVER PROTECTION

Titanic housed an office for the wireless company The Marconi Company. The wireless allowed the passengers to send and receive cables, a luxury in communications at that time. It also allowed other ships in the area to transmit the coordinates of ice flows. Policies on these types of cables were simple: the cables with the locations of ice fields or icebergs were given to the bridge officer, who would then tack these paper cable transcriptions on to a board in the bridge. This provided current information to the officers or those coming on the next shift.

The Marconi officer working the night of the sinking logged that he gave a cable to Capt. Smith in the early evening. He remembered that the Captain was in a rush and that he placed it into the right-hand pocket of his jacket with just a precursory glance. There were six iceberg warnings received by the *Titanic* on the day of the disaster. One message got to the Captain, but the others were ignored by the wireless operator who was busy transmitting passenger messages. Passenger messages were the Marconi's bread and butter. The wireless was clearly less available for operational messages (which generated no money for the company) because of the substantial profitability of the passenger use.

NEW RULES OR OLD RULES?

Company policy only required the wireless offices to be operating until midnight. *Titanic* hit the iceberg at 11:40. Most of the ships in the area had turned in for the night by the time the ship was transmitting calls for help. For several years the distress call for ships was CQD, which translates as: CQ = attention all stations, D = distress or danger. A short time before the *Titanic's* launch, an international convention introduced a new distress call to supersede the traditional CQD. The letters chosen were SOS, not because they stood for anything in particular (although rumored to be Save our Ship), but because they were simple enough for the new amateurs flocking to the wireless craze to send and receive. *Titanic* was tapping out at 12:15 "CQD MYG" (the ship's call letters). *Titanic* was the first ship to have ever used the newly negotiated distress call SOS. The ship's wireless crew tapped both CQD and SOS in order to cover all bases.

At 12:19, on the Cunard Line's *Carpathia*, just as the wireless operator prepared to turn in that night, he heard the call. The ship was less than 60 miles away. The *Carpathia* picked up and saved the *Titanic* survivors. Had the operator turned off his wireless, the distress call would not have been heard and no rescue craft would have responded.

At around 10:30 pm, the liner the *Californian*, a cargo ship, had stopped for the night on the edge of an ice field. As was the custom of the time, the wireless operator had turned the radio off and had gone to bed. The ship's night crew noticed a behemoth liner stop an estimated 6 miles to the south at 11:40 pm.

Shortly after midnight, the Captain of the *Californian* was informed by the crew that an enormous passenger liner was shooting rockets into the sky. The crew concluded that this ship had anchored for the night and that there was some type of party. These were not fireworks but *emergency* flares. At 2:20 am, it was noticed on the *Californian* that the *Titanic* had disappeared. Those on the Californian believed the other ship had steamed away. The *Californian's* wireless operator woke up in the early morning and turned on his Marconi radio. It was at that moment the crew learned of the tragic fate of the *Titanic*. In both the British and U.S. Senate inquiries into the disaster, the Captain of the *Californian* insisted his ship had been many miles north of the *Titanic*, and that he could not have reached it in time to rescue passengers. However, many survivors testified in the investigations that they had seen the lights of another ship roughly six miles north of the *Titanic*. Both international inquiries concluded that the *Californian* may have been only six miles or so to the north of the *Titanic*. The *Californian* could have reached the *Titanic* before it sank.

DISCUSSION

Since you have now read this case study, again ask these three questions:

- Whose fault was the sinking of the *Titanic*?
- What caused the *Titanic* to sink?
- What corrective actions would you have recommended for the next ocean liner to be built?

Remember these questions and your initial answers. Have you reinforced your initial answers, or are you now reconsidering your initial answers?

The tragic story of the *Titanic* is a story about so much more than an iceberg. It is easy and profoundly wrong to assume that an iceberg sank this ship. It is also profoundly dangerous to think that an iceberg caused this event, because focusing only on the iceberg takes the safety learning attention and the allotted corrective action resources away from the real weaknesses and potential learning opportunities of this event. Worse yet, blaming the lookout or the captain, although a tight story and emotionally satisfying, is not terribly accurate. No one person alone was able to have this failure: this failure was a product not of the crew, but of the relationships between the crew and leadership, the crew and passengers, and the designers of an unsinkable vessel.

No one person or thing caused this failure. Instead, a whole series of small, seemingly innocent, normal activities combined in such a way as to cause the safest vessel ever built to cruise the seas to sink. No mechanical reason alone caused the ship to sink. However, the mechanical watertight cell walls not going all the way up to the above boat deck is, in retrospect, a failure. However, had there not been doors in the cells, water would not have been able to move from cell to cell. If there had been no doors the crew and passengers could not have moved throughout the ship during normal cruising.

Interestingly enough, no one error caused this failure either. In fact, there is an argument to be made that in almost all cases, no errors were made in context. The errors only became apparent after the *Titanic* sank. Confusing emergency flares for party fireworks was not an error until the realization was made that the ship that was six miles away having a party was actually sinking. Even the captain placing the ice flow Marconi transcription in his pocket would have had little effect if the other five transcriptions had been completed.

All of these small, unpredictable components of this failure seem very important and very dependent upon one another when you think of how the *Titanic* failed; however, in the context of operating an unsinkable, technologically advanced ocean liner on her maiden cruise, no one event would have come close to triggering even a low level alert. It is not until after the failure happens that these seemingly normal deviations from expected outcomes become very important warning signs.

This is a complex (many parts, many people, many systems, many stakeholders, many goals, many pressures—all closely related to one another), adaptive (decisions being made in real time based upon information that is readily at hand in order to create the best outcome at the time) failure. No one cause. No one person to blame. No errors to discipline or train away.

3
Change the Way Your Organization Reacts to Failure

QUESTION ABOUT YOUR ORGANIZATION:

How does your organization act after something unexpected happens?

1. We have an immediate meeting and decide what happened, and how to fix it.
2. We pull the worker(s) involved into a meeting about the event.
3. We name, blame, shame, and retrain.
4. Somebody's going to get a phone call—probably me.
5. We see it as a chance to learn and probably will compile a team to learn as much as we can.

A STORY

This is a story about a maintenance man, and a small carnival operation headquartered in the Midwestern part of the United States. The carnival traveled from town to town, county fair to county fair, with 20 amusement rides, 6 dark rides, and several games of both skill and chance on the Midway.

The equipment in this carnival was aging, and demanded almost constant attention from management, ride operators, and the maintenance man. There was enormous pressure, because of the small, mom-and-pop size of the carnival, to keep every amusement ride in running order. After all, a broken ride cost the carnival company money, and did not make the carnival company money. In order for the maintenance man and the rest of the carnival workers to get paid on Fridays, all the rides needed to run as full as the can all week.

Times were hard. It seems that not as many small towns had reasons to have carnivals, and bigger carnival companies with better, more exciting rides were picking up the state and county fair jobs. The pressure to find work was high. The pressure to delight the customers became more intense. The problem was that while performance pressure was increasing, resources to ensure the expected level of performance were declining. Times were hard and seemed to be getting harder.

Several weeks previously, "The Whip" had started making a strange grinding noise. "The Whip" was a high speed swing ride, where the riders sat in small cars suspended by cables from a central spinning motor axle. The ride was designed to start slowly, then turn to the right in a circular motion, eventually turning faster and faster. When the speed neared its maximum rate, and due to the nature of centrifugal force, the little cars filled with riders would be traveling at about 20', parallel with the ground, for about 4 minutes.

To say that "The Whip" was exciting is probably an understatement. "The Whip" was fast, fun, and dangerous. Everything you want from an amusement ride. But the ride started to make a grinding sound; the sound was an indicator that something was not right with this ride. Shutting it down was not the best choice. The ride was one of the carnival's more visually appealing rides. The carnival company had evidence that something was wrong with the ride, and had to make a safety decision with operational consequences.

Anyone could stop the ride if they felt it was unsafe, but a little grinding was not unusual, and the line of riders had been long and steady all weekend. "The Whip" could be stopped all right if the grinding got much worse—the problem was there was no standard for what worse grinding sounded like to each operator.

So the decision was made to perform maintenance on this ride on Saturday night. As soon as the fair closed, the maintenance man would troubleshoot the ride, find the problem, and fix it. With any luck, the entire ride would be up and running and full of riders by 10:00 am Sunday morning, when the fair opened for its final and most profitable day.

The maintenance man tore the ride down, and checked everything he could think of checking in order to fix the ride. The more he worked on the ride the better the ride sounded. It was almost back to working order, and seemed to be operating at about 90 percent. The maintenance man did not have time in one night to take apart the transmission housing, and check the main linkage between the transmission and the ride arm axle. That process was easily a two-day job. He would hold that task until Monday and Tuesday. The beginning of the week was the time when the carnival either moved, or in rare cases gave their workers a couple of days off. This partial repair would have to do—it wasn't perfect, but it was good enough to make it till Monday.

At about 6:30 am Sunday morning, the maintenance man tightened his final bolt, and turned the ride to the operational setting. He checked to make sure the operator's logbook noted the repair, and did the paperwork to make the ride operational for that day's riders. The maintenance man then went back to his travel trailer and went straight to bed. He was getting too old for this crap, and he was certainly feeling it when he had to do an all-nighter. Sleep would be welcome, and come easy.

But it didn't, not at all. In fact, the maintenance man could not shut off his brain. He just kept thinking was "good enough" really good enough? Many lives would depend on this ride not failing, and the maintenance man could not say for certain that the grinding was not a sign that the main coupling was getting close to failure. Had he greased it enough over the past couple of years? Was the grinding a sign that it was wearing badly and may be close to breaking?

On Sunday morning, at 9:15 am, the maintenance man got back out of bed, went to the management, and told them what he was thinking. He talked to his bosses about the potential of the coupling giving way and literally throwing riders across the fairground at 40 mph, at an altitude of 20' above grade. It was too risky, and he just did not feel right about his work. The carnival must cancel operations of the most exciting, highest grossing ride on the biggest day of the fair.

And so the management canceled the ride. The ride manufacturer was called, and sent a field inspection team to do an emergency inspection of the ride. The manufacturer's field inspection team went ahead and pulled the transmission housing off the ride, and discovered, to everyone's worst fears, that what was once a 2½ inch-thick steel axle had worn to less than a quarter inch. Had the ride operated that day, *that day*, 50 people would have been seriously hurt or killed, and the carnival company would have gone out of business.

In essence, the maintenance man, and his ability to gather the nerve and fortitude to call stop on "The Whip," saved the lives of the riders, his carnival company, and perhaps the carnival industry. The maintenance worker was given a $1000 bonus, recognized by both the county and the state for his heroic efforts, and featured in the local newspaper and even on the cover of the professional magazine of carnival operators.

All in all, the worker and the company together did exactly the right thing, and safe operations saved the day. The world was safe, reliable, and productive once again.

But, just for a moment, let's play out another ending...

What would have happened if at 9:15 am on that same Sunday morning, that same maintenance man had told that same story? The ride would have been shut down in the same way. The manufacturer would have been called in to the fairground. The manufacturer's field inspection team would have flown out at the carnival's expense to the fairground in that small Midwestern city. However, imagine that this time when the field inspection team took the housing off the transmission in order to inspect the coupling, all they found was nothing. The axle was perfectly greased, and the noise was simply normal. The grinding sound was well within the operational parameters of the ride's specification.

Would the maintenance man have been rewarded? Would he still have received the $1000 bonus? Would the county and state have recognized him? Would he have received any praise at all?

...Or would he have been seen as overly careful, foolish, and not a "smart" decision maker?

DISCUSSION

This is an interesting story. How you react to failure, or react even to the *potential* for failure, matters to how safe your operations are operating. It is easy to celebrate a worker who exercises good judgment with a questioning attitude when the worker's questions lead to a discovery of a potential failure before it becomes a failure with serious consequences.

It is not so easy to celebrate exactly the same behavior and decisions when the outcome is not a potential high stakes failure, and, in fact, costs the company a ton in lost revenues. The natural inclination is to say that the worker was too cautious, too careful, and overly vigilant— and all of those things are bad if there is, or was, no threat of failure.

And yet we are talking about exactly the same behavior, the same attitude, and the same worker who worries about a problem and reports the problem to their management. The decision to stop operations was a costly one, but when it avoided a real problem, a potentially horrible accident, the cost went from being a burden to a blessing.

In many ways, this story illustrates much about many organizations. You must first understand the many inputs that go into your reactions to events. It is not an easy task. It is sometimes costly, but ask yourself this: if you were the maintenance worker, and the story took the second ending—"nothing wrong here"—and you were "punished" for being overly cautious, would you ever report anything to your manager again?

Your biggest challenge is to create an organization that is not afraid to talk to each other about safety. This is no small challenge. To do this you must first understand that to expect trust, you must give trust. To expect communication and a questioning attitude, you must build an environment that fosters those behaviors. To do that, you must always think of how your organization reacts and responds (two remarkably different words and ideas) to failure.

WHY YOUR REACTION MATTERS

Something unexpected and bad happens...

...And everything that you and your organization does from this moment on will become necessary to creating everything else that happens—the critique or after action meeting, the investigation, the corrective actions, the whole response picture. Time matters, because it is impossible to have an organization that is operating out of control.

This situation is tough, because your first reaction after an event has been to identify who made the mistake and think immediately about ways to fix that person. The old view of accidents was a pretty narrow view—simple, sweet, and cheap to fix. If you find the guilty person, and remove or retrain that person, you have clearly fixed the problem. Here is a problem with that old model: the "bad person" model tends to lead any organization down the linear, mechanistic path. The problem with this path is that it feels as if you have done something, have fixed the problem, when in reality that could not be farther from the truth.

It seems simple, almost too simple, but your organization's reaction to a failure tells your workers everything they need to know about your company, its managers, and what is going to happen. How you respond sets the stage for everything that will happen, and that has happened, and for how the event happened in the first place.

I am going to say something that you should question immediately. In fact, in normal circumstances I would never trust anyone that even came close to saying what I am about to say. If there ever was a time to call "B.S." on anyone, this would be that moment. Here it goes:

> I know how you can change your organization's entire safety, quality, security, or production performance for the better, immediately... Change the way your organization reacts to its failures.

It is a bet: if your organization is anything like my organization, when something bad happens your organization jumps almost immediately to some type of faultfinding posture. Who screwed up this time? The new philosophy I am proposing helps your organization move from the classic "crime and punishment" reaction to a "diagnose and treat" response to failure. That shift is exactly what we are hoping to create in the new organizational view of safety.

TOOL: BETTER QUESTIONS, BETTER ANSWERS

You must help your managers know this little but important fact. You must tell your managers that the way they react or respond to an operational failure matters—and it matters a lot. Coach, counsel, and press your leaders into knowing that they play an enormous role in event learning, and organizational culture.

All of this boils down to the premise that your mangers will need to know what to do. If you want them to do something differently, you must tell them what that different thing is that they should be doing. Hence the following job aid. This is a card that we produced which asks the managers to be aware of both their reactions and responses, and suggests some better alternatives to ask when confronted with an event.

Nine Things Managers Should Ask First and in Order When They are Notified that Something Unexpected Happened:

1. Your response to an event matters!
2. Are the people OK?
3. Is the facility safe, secure, and stable?
4. Tell me the *story* of what happened?
5. What could have happened?
6. What factors led up to this event?
7. What worked well? What did not work?
8. Where else could this happen?
9. What else do I need to know about this event?

At my organization, we put these nine questions on a card and handed that card to our managers. I discussed this concept earlier in the book, but here is what my organization found. Our managers want to ask the right questions; we had not done a decent job of telling them better questions to ask. Interestingly, when our managers actively tried to gather explanations and stories, and not reasons and causes, we started to get decidedly different, much higher quality reporting of events.

With a better understanding came better responses and better solutions. With better solutions, we saw organizations drop not only the number of their events, but also the severity of events.

CRITIQUES, AFTER ACTION REVIEWS, FACT FINDING MEETINGS

> "This meeting is being called to understand how we had an accident in area 7 this afternoon. This meeting is not about blaming workers. We want to have an open conversation about this situation. Feel free to communicate openly and honestly to management...Now, tell me how you guys screwed this up this afternoon."

Actual Critique Opening

Your facility most likely holds some type of event review meeting after a failure. In larger facilities, this meeting can be quite serious. Smaller facilities may not have the same formality and size of this type of meeting, but it is almost certain that this type of meeting happens in these facilities too.

These meetings are usually used to perform several critical tasks. The primary reason for these meetings is to understand what went wrong and decide what needs to be done, usually at that very moment—or moments after the meeting. The second reason to have these meetings is to categorize the event by some standard. Is the event in question a really an event at all? Is the event a product of bad worker choices? Is this event a result of management actions? Does this event fall someplace in between these two extremes?

But here is the problem with these after event meetings held after something terrible happens. These meetings become a little bit like a court hearing, and often not much of an opportunity to learn "how" the event or events transpired.

Here is a question you should ask about your organization:

> "When was the last time you had a full blown critique after a job was completed successfully?"

It is easy for your organization to find resources to understand and investigate a failure. We always have the time and money to do an investigation. In reality, because it is an event, we know that we will have to pay for some type of learning activity. The problem is that it is not easy or even practical to understand and investigate success. Success is normal. Success is common, and, therefore, we don't actually see a need to understand why a job goes well. Even if we did track successes with disciplined regularity, how would we find a data management system to keep all the success data that we would find?

Post-job reviews after operational success are an excellent opportunity that is often missed or wasted by organizations. These meetings ought to be an opportunity, open and honest, to tell the whole story of the success that just happened. Workers should feel that these meetings are fair and open, and that the management is genuinely interested in learning about all the factors that made up the success. Looking at failure with the same type of meeting you use to view success sends a strong message to workers and managers. The message is that your organization is interested in learning no matter the outcome.

Often the room itself is an indicator of what the meeting is like for the workers involved. You might be surprised to know that the management conference room is not the psychologically safest and most open place for the workers. Many times I have moved chairs and tables, brought in coffee and doughnuts, and asked senior managers to "non-attend" these meetings in order to create a space where the workers may speak openly and honestly.

If you are interested in building a culture in your organization that is able to report events in ways that best tell the whole story about a failure, you must be interested in changing this very formal after failure activity. It is helpful to be open to the reality that these meetings, which often feel truly open and safe to management, may give a completely different impression to the workers who are required to attend.

TOOL: MAKE THE FORMAL LESS FORMAL

Try these three things the next time you have an opportunity to have this type of meeting:

Don't just start the meeting telling the members of the meeting that this will be an open meeting, nothing will be used against the worker, and our goal is to learn—do something to demonstrate that this meeting is different. Work at doing whatever you need to do to make this learning opportunity feel and look more like a chance to gather information and less like a courtroom proceeding.

Make this meeting less formal, less "courtroom-like," and more about a chance to go about honestly learning about the story of the failure. Use the "9 questions managers should ask" when notified of an event in your organization to guide this post-event meeting.

Have a post-event meeting for work that was done successfully, and invite lead workers and opinion leaders, and ask them to tell you the story of what went right, and why we as an organization were successful in completing this task. Take this meeting seriously, and go into this activity with the idea that you want to learn about what works as much as you want to learn what went wrong.

Be aware that reactions to events tell workers everything they need to know about your company's safety program. If you do not attempt to "fix" this problem, everything, seriously, is communicated about how your organization will move forward. It is more important to focus on how your organization will learn from this event and avoid future occurrences.

Learning from success will surprise you. Successful completion of work in your organization includes just as much worker judgment and work improvisation, and just as many adaptive behaviors and surprises as work that fails. You can learn all the same stuff you learn after a failure, without the costs of the actual failure.

This whole process is exactly how we make safety better in our organizations.

TOOL: LEARNING TEAMS

One of the better tools my organization developed to increase and improve operational learning is a remarkably straightforward tool with a remarkably high level of effectiveness. This tool was developed to address the need we had to make event learning more effective for our facility. We knew there was lots of information on our job sites, but we were not getting much operational information. We had an immense challenge, and the challenge was this: we needed a way to engage our workers in event learning and a better way to get important information about our systems and processes.

We knew that we could achieve this if we could just tap into the informal learning that was taking place at the task level. We knew the workers always seemed to have a much more complete contextual understanding of how failures and near misses were happening in the field. We needed to capture this information in a way that reinforced and respected the workers' story, while at the same time was an effective communication tool with, and for, our management.

What we came up with was the learning team. A learning team is an ad hoc group of workers from any level within the organization that have been brought together to answer one question: Something has just happened— what should our organization learn from this event (or potential event)?

These teams, or, better said, this idea of being able to create a team at a moment's notice, could be called upon by a manager at any level of our organization for problems (or potential problems) that warranted a little bit more attention. However, this type of operational learning also needed to be legitimate enough to use as a part of event response for the regulator.

In your guidance to learning teams, you should tell the team members that you are not interested in "who" did what (we almost always know that information) or in "why" the event happened ("why" tends to lead to a more mechanistic failure picture); you are most interested in "how" the event transpired. You desire the story of the failure—with a clear beginning, a clear middle, and the ending. This can't be stressed enough: *how* is more important than *why*, and *why* is more important than *who*.

Once the team starts building the story, you ask them to produce some type of report or product. This could be a set of slides, a written report—whatever the team thinks is the best way to communicate what they have learned. You then have the team report this information out to some appropriate level of management.

If this all seems rather unstructured and loose, then that is because it is unstructured and loose by design. In your organization, you want a learning team to be a management tool that exists somewhere before a critique and well before an investigation. This tool exists not to create a formal record or some type of auditable document, but instead is a tool for the organization to learn and correct from. You also want your learning teams to be a worker-owned process that provides a way to encourage worker ownership and to gather worker-level information about operations.

The challenge has been in trying to capture these learning team reports at the corporate level. I have found these teams incredibly valuable at the local level. It is my guess that this information will have potential value for the entire organization. At my organization, we are trying to crack that code even as I write this information for you.

The challenge persists that as this process is formalized, you will lose some of the honesty that you are getting by using this process. If this process becomes formalized and structured, the process itself will move from a way to gather real, leading safety information to a lagging deliverable to either the organization or to the regulators.

Try this method and see what you find. Customize this tool (and others) to fit your needs. If you communicate and manage these learning teams effectively, which is pretty easy—leave them alone and give them broad expectations ("tell us what we should learn from this event")—you will soon discover that learning teams become your corrective action planning program. Workers know what keeps them safe. Workers keep your organization safe.

Pre-Accident Investigation Tool

EXPANDED TOOL DISCUSSION

Have you ever watched one of the many CSI (Crime Scene Investigator) shows that have been on television recently? These shows tend to feature an almost "magically logical" deconstruction of a crime in order to track and capture the bad guy. It is done neatly in 60 minutes, and often involves what we have called in this book a more "mechanistic" investigation of the crime for most of the show, until sometime near the end of the hour the CSI team puts all the information together in a more holistic way in order to tell the story of the crime that happened. The story is usually a complete description of how the crime happened, what the motive was for the crime, and, most importantly for the CSI show, who is the culprit. The concentration is not only on who, but the why, what, when, where, and how of the crime.

In essence, the CSI shows are exceedingly powerful examples of the Human Performance view of understanding failure. The investigators strive to understand and explain what happened without judgment, in order to understand the story and to provide a just and honest conclusion to each case.

A just, honest, and concise explanation is precisely what every Human Performance investigation strives to provide. We look for the story of the story: a complete, context filled explanation of how the event happened is the best possible outcome of any formal method of understanding failure. Often the Human Performance investigation gives the organization information that is incredibly comprehensive, makes it easier to identify what to correct than with old school methods, and actually helps build a culture of trust and communication.

The only problem is that in order to get this CSI-type information you have to have a failure. There is nothing good about having an event in order to learn how not to have the very same event. Or is there? We are pretty sure there is.

We can often get the story of the failure before it becomes a failure. It is simply a matter of time, access, resources, and luck. Sometimes we discover problems in waiting. Other times we are surprised by events that we wished we had discovered beforehand.

Here is one way to help formalize and structure your prevention strategies by leveraging your organization's ability to learn. Think of your pre-accident investigation as a 6 part process.

1. Look for high consequence activities
2. Look for small signals that can indicate system weaknesses or problems within the normal work process
3. Look for error provoking systems steps and processes
4. Look for error likely conditions
5. Listen to your workers
6. Ask yourself what keeps you up at night?

Identifying Pre-Accidents in Your Organization

You can't fix everything. You don't need to fix everything. In fact, you run the risk of making things worse if you run around trying to fix processes, procedures, and systems. Yet, you need and want to fix the systems that are setting workers up to fail at your facilities.

Here are some quick guidelines for identifying potential areas for systems improvement in your organizations:

Look for high consequence activities
These are the tasks you know will cause consequences if these processes fail. You know these tasks involve many high risk, high consequence factors as a normal part of doing the work. It is fair to say that these systems normally work well, and either rarely fail or have never failed. These potential failures can, and will, cause unacceptable consequences for your organization, your people, or the world at large.

These specific tasks either have high risk hazards or high value materials. These systems often have remarkably robust safety programs associated with them. If you are not assessing these programs for the presence of multiple layers of defenses and operational surety, you should start today.

You probably aren't looking at these systems and asking, "When this process fails, what safety defenses will reduce and control the consequences of this failure?" These systems don't need a stronger industrial safety program. You are probably sure that they are quite reliable, as the process exists now. However, what these processes demand is an aggressive systems safety program.

Assume the failure will happen. Then when the failure does happen, you will be prepared for the failure and its consequences. If the program does not fail, you will still be ready for failure to happen.

Look for small signals that can indicate system weaknesses or problems within the normal work processes
Errors, near misses, good catches, close calls—any of these factors could indicate there is a problem, without the actual consequences of the failure.

Safety professionals look on indicators of this type as "gifts." There are many examples of these small signal events happening every day in your organization.

You must monitor low level events in order to understand where your systems are confusing, conflicting, or potentially flawed. Don't monitor these small events because they are mathematically predictive (the more these events happen, the higher likelihood that we will have a serious event); these small events aren't truly predictive.

What these small events do is to allow an organization to "pulse" its processes and systems to identify potential larger system weaknesses or failures. The premise is that these low signal indicators are in reality pointers towards larger system weaknesses and failures.

Look for error-provoking system steps and processes
These types of conflicts are all over your organization. You know these situations: the cases where if the worker were to follow the process the worker would certainly fail. These are the conflicts that arise in any organization's systems that place workers in positions of uncertainty—while the organization assumes that there is clarity of decision. The classic example used for this discussion is about the difficulty of using a purchasing system.

For example, you need to buy a screw for a project you are doing in your job. The screw will cost about 50 cents at a hardware store. To use the company's purchasing system will take three signatures, filling out a couple of forms, and about two hours of time. What is going to make you use the system when you drive home by a leading lumber seller every day? It is actually much easier, faster, and less painful to buy the screw with personal money than to use the organization's systems, and the organization's money.

Look for error likely conditions
Error conditions are everywhere. Look for the space between your workers and the work they are doing for conditions that could cause failure. Identify places where failure could happen because of the worker simply performing work. They are everywhere. Better yet, look for well-designed systems and conditions that direct workers away from error and toward success. You will not be surprised to discover that many of your systems actually do set workers up for failure. Many of our systems are set up to be effective accounting systems, and not good work management systems. These systems are rife with potential failures.

Listen to your workers
Ask your workers where the next accident will happen; you will be surprised by what you will learn. They are brilliant at this CSI task. Your workers know where your system makes sense, works well, and is efficient. They also, by definition and practical experience, will know exactly where the system is setting workers

up to fail. Remember that when you ask workers for this type of input, you must then do something with this information.

Knowing what the workers believe to be failure-prone becomes a contract with the workers to improve this situation. The key concept here is that mitigation of this type of problem is often better and faster than long-term, expensive solutions. It investigates potential failures before they could take place.

Caution is noteworthy here. If you ask workers to tell you about potential failures, you must be prepared to hear honestly, accept, and act on this information, and not justify and protect systems over people. Don't defend the process over the opinion of the workers. They are your frontline experts.

It is no secret: if you want to know how work actually gets done, ask a worker what is truly going on there. They have the day-to-day practical experience.

Ask yourself (and your peers) what keeps you up at night
Like listening to the workers, you are now creating another path towards capturing another perspective of your work environment. This question is as strong as any other tool used to predict potential failure. Have your fellow managers and supervisors tell you what they think is the most risky activity (or part of an activity) on the shop floor.

Remember that risk perceptions change the closer you get to the actual work. In most cases, risk perception decreases the closer you get to the actual work. So, by definition, management's perception of risk is going to be different to the workers' definition of risk. Managers may be away from the work, but what they don't have at the worker level they most certainly make up for at the systems/relationship level. Your fellow managers know how work fits together: the bigger picture. These people can talk about pressure, resources, and worker personalities.

Get real information
Small things that go wrong are most often warning signs of growing trouble deeper inside the organization's systems and processes. This type of data helps to provide remarkable insight into the health of the entire organization; almost everything that goes wrong is some type of indicator, some type of data. Look deeper into the context of what small failure can mean. Failure often has subtle beginnings and plenty of retrospective indicators.

Near misses and close calls are a function of how much your workers trust you and your organization. Don't underestimate how many close calls your organization may be experiencing every day you are operating. The bigger question here is how many close calls and near misses do you know about, and how do you handle the low level events that actually get to your radar screen?

Where you are in the organization assuredly affects the type and the amount of information that you get from the field. For example, a senior manager by definition will have the view from the top—even if the senior manager believes he or she understands the workers' perspective. Just by entering a workspace as a safety professional, you change that workspace.

The goal is to identify and use multiple ways to get honest information from workers. Once you feel you are getting honest feedback, accept it. Process the information that you get—not the information that you wish you were getting—and don't be defensive. It is easy to try and explain what is going on when you receive a worker comment, but don't do it. Accept the honest information deliberately and thankfully. Remember, the worker is giving you a chance to see the work from their perspective. Actually, that is almost a sacred moment and should be considered significant and vital.

Getting real information has more to do with trust and relationship building with those closest to the potential failure than it has to do with record keeping, computer systems, and accounting. Worry how you will manage the information after you figure out how to build trust levels to a place where your workers can speak honestly about systems and processes. Then take this data and act on it.

Interview your workers

Ask workers what works and what does not work within your organization, and then listen carefully to how they answer these questions. Where do your workers believe they are the safest? Where do your workers believe they are least safe? How do your workers create safety while they do their work? What hazards are they skilled at identifying and mitigating, and which hazards are they surprised by? When do your workers have to adapt and improvise in order to do their work? This is a gold mine of data.

Just as you would if an event had taken place, you would bring in small groups of workers and ask them to tell you how work actually happens. You know how work is supposed to happen, how the system was designed to take place, but do you know how the work actually happens? This interview is to help you capture the story of the work that is being done. In capturing the story you will learn of the parts of the work that are safe, and the parts of the work that need your attention.

Mostly, you are interviewing workers to ask them to tell you the story of working in your organization. This dialog will tell you where failures can happen before they happen.

Investigate the accident that hasn't happened…yet

What you go out and look for is what you will go out and find. Be careful. You must be acutely aware of how your pre-investigatory biases will directly impact what you will find. Since there has been no event, the normal "smoking ashes" that you would sift through are not present in a pre-accident investigation. This forces you to be imaginative and creative in the way you proceed in this learning opportunity. You are honing your CSI skills.

In short, what you want is not an investigation, but in reality an explanation of how failure would happen, if it were to happen, in the work being done by workers. You are writing a story of what could happen if the organization did not intervene in the process. A story is a powerful tool because it has a beginning, middle, and an end. A story must make sense, be plausible and realistic, and move logically through time and space. You can sensibly explain what could happen, and then how to prevent it.

This is the time you will capture information at a detailed level. The first thing you will want to do is to start building a timeline. The difference between this timeline and an actual failure timeline is that your timeline for a pre-accident investigation does not have an accident on the right end of the chart. What this timeline has is an intervention and defense placement on the outcome end of your chart.

Never look for cause and effect outcomes, instead look for relationships between workers, workers and management, and, most importantly, workers and your systems and technologies. Looking for these relationships forces your organization to understand and "deal with" the context of your work. It is within this context that you discover the potential failures before they become failures.

Now, engage your learning to strengthen your systems against the potential failures you have uncovered

You're not ever going to be able to stop an accident. You can directly change the way the accident affects your organization, your workers, and yourself. A pre-accident investigation helps you make your organization better prepared for a failure. Our message is consistent and straightforward: "We can't stop accidents, but we can prevent consequences."

Engage your learning systems to make your organization smarter and more prepared for the potential failures you uncover.

You are gathering information to prevent adverse outcomes for your workers and your organization. You will never be able to measure what doesn't happen. You will never be able to predict every event. Yet, it is clear that if you can gather enough information about a system to identify the places where failure is most likely to happen, or places where if a failure were to happen it would have some type of serious consequence, you can actually intervene in your organization's processes and systems, and prevent events.

You aren't actually looking for the next accident. You can't ever predict the next accident, because these events are always surprises to the worker and to the organization. What your job becomes is to identify places within your organization where the organization sets up workers to fail. Look for conditions that lead to failure, not trends or numeric patterns, and when you find these conditions, learn from them.

It all boils down to how your organization, your managers, your teams, and you learn from your organization's normal, routine work—day in and day out. The only tool you have to prevent events from happening is your organization's ability to learn as an organization—to learn honestly from itself. Use your learning systems to serve the greater good, not to function as the accounting method to document your safety history.

You will struggle to look for indicators that you can use as predictive. Every organization struggles to identify indicators that can be used to predict the next critical event. I am convinced that there is no such thing as a *leading* data set that can help safety professionals predict the future. I can make this claim because if this data set were out there, I am positive that someone would have found it, and we all would be using these numbers and methods.

Most importantly, ask people what conditions we should identify. Where are we most prone to having a problem? Where are we sending conflicting, multiple signals? Where is work overly controlled? Where is work under-processed, under-resourced, or misunderstood? Ultimately, use your information to the best of your ability.

How will you know you are correct? How will you know you have been effective? The quick answer is that you never will know how many events didn't happen because of your pre-accident investigation process. You will never be able to measure something that does not happen. Be satisfied in the process, not in the outcome. After all, if you have done your job well, nothing unexpected will happen. What a beautiful thought for your workplace—imagine, a workplace where nothing unexpected happens every day.

Case Study
Aviation Accidents are the Unexpected Combination of Normal Aviation Variability

"When everyone is to blame, is any one person blameworthy? Does blame have any value at all in better understanding this event?"

This case study demonstrates how almost no indicators of failure are present before the deviation from the expected outcome transpires. Perhaps even more interesting is the rather normal nature of all of the factors involved. No one factor seems important alone. No one indicator is even considered operationally interesting by itself. Had there been any indication, at any level, that this outcome was even possible, many different decisions would have transpired on the night of December 20, 2008. It typifies a difficult conundrum of safety management: no predictable indicators of failure *before* consequence happens.

While reading this case study, observe how this event is made to appear linear (it is actually the only way to talk about this event in retrospect). Contrast this same linear failure with the actual failure that was the outcome of multiple conditions all coinciding nearly at the same time. You will see how these conditions that are cited in this event all seem to happen repeatedly within the telling of this story. It is important to note that this case study is the product of an aviation insider telling a context-rich story of a successful accident.

Finally, notice how the absence of any formal operational leadership (due to injuries sustained by the captain and first officer) after the event gave rise to the opportunity for leadership to move throughout the event's context. Leadership for event recovery, event response, and passenger evacuation moves many times throughout this event.

THE BURNING OF FLIGHT 1404

On Saturday evening, December 20, 2008, flight #1404 crashed on take-off from a runway at Denver International Airport, bound for George Bush Intercontinental Airport, Houston, Texas. The accident resulted in many passenger and crew injuries, and a hull loss of a Boeing 737-524 aircraft. According to the preliminary information provided, the plane veered off the runway early in the take-off roll, ultimately crashing into a ravine approximately 200 yards from the runway center line and erupting in fire. The aircraft sustained severe damage. The fuselage was cracked just behind the wings. Engine number one and the main landing gear were sheared off. In addition, the nose gear collapsed. The fire caused overhead luggage compartments to melt onto the seats. The fire was on the right side of the aircraft. All on board evacuated through the doors on the left side of the aircraft.

The initial emergency dispatch call to the Denver Fire Department was incorrect in the location of the aircraft. However, despite the early confusion as to the whereabouts of flight 1404, firefighters were on the scene relatively quickly, as the aircraft came to rest near one of the airport's four firehouses. When the Fire Department arrived, most of the right side of the plane was on fire. The passengers were evacuating from the left side, assisted by flight attendants and one off-duty airline pilot in the passenger compartment. This same pilot made several trips in and out of the wreckage to ensure everyone was safely out of the aircraft.

Initial evaluation of the accident cited rudder issues as the reason for the aborted take-off. Further investigation indicated issues involving a combination of high crosswinds, or wind shear, and the presence of aftermarket winglets installed on the wings. An additional reason was the captain's inability to keep the aircraft on the runway during take-off. A plane that leaves the runway during take-off is said to have had an runway excursion. Try to keep all of these reasons in mind as you sift your way through this case study.

This was the most serious accident in the thirteen-year history of Denver International Airport. Luckily, there were no fatalities. However, there were some notable injuries.

A SUCCESSFUL ACCIDENT?

Flight 1404 was scheduled to depart Denver for Houston at 18:00. The flight crew arrived at the airport about 17:00. The airplane had not yet arrived at the gate. The first officer bought coffee while the captain walked downstairs to get flight paperwork from an operations coordinator.

The flight crew consisted of two pilots: a captain and a first officer. The day of the accident was the fourth day of a four-day pairing for the two pilots. The pilots had been paired with each other on one or two prior trips. The last trip had been about a month before the accident. The pilots reported a history of positive, professional interactions. It was a typical day and a typical flight.

The pilots met at the gate after the airplane arrived. The captain did an external preflight inspection, while the first officer performed preflight safety checks in the cockpit. When the captain returned to the cockpit, the two pilots discussed the upcoming flight; performed the receiving aircraft checklist; obtained a passenger count, and weight and balance information; and entered the load information into the airplane's flight management computer. All these are standard preflight tasks.

The flight crew instructed the cabin crew to close the airplane's doors. The first officer contacted ramp control and received a clearance to push back from the gate for a west taxi. The flight pushed back at 18:01. Ice and snow were visible on the ramp. Following standard practice, the captain started both engines, and turned on the engine and wing anti-ice systems. The first officer contacted ground control, and received a clearance to taxi to runway for their departure.

The flight crew heard ground control tell the flight in front of them that Automated Traffic Information System (ATIS) "Sierra" was current. Winds were reported at 270 degrees and 11 knots (about 13 mph). They continued to taxi on the taxiway toward the runway. The captain did not notice any buffeting of the airplane from wind during the taxi.

As the airplane approached the runway, the flight crew performed the before take-off checklist and contacted the tower. An additional aircraft was on the runway ahead of them, awaiting a take-off clearance. When the awaiting plane departed, the tower instructed flight 1404 to position and hold on the runway. The runway appeared to be clear of snow and ice, so the captain decided to deselect the engine and wing anti-ice systems, but he left the engine igniters on. The captain positioned the airplane on the runway, and the flight crew waited for two or three minutes. The cabin crew waited in their seats. The passengers were thumbing through magazines, adjusting pillows, and talking amongst themselves.

The runway lights and all of the airplane's lights were operating. Visibility on the runway was excellent. The captain later recalled that the airplane's speed was probably less than 100 knots (115 mph) when this event occurred, as he had not yet heard the first officer's 100 knot callout.

The captain received no system warnings and no signs of obvious system malfunctions. He kept the right rudder fully engaged as the airplane veered toward the edge of the runway. His right hand remained on the throttles and his foot continued to depress the right rudder pedal during this time. He did not touch the brakes because he did not want to interfere with the auto-brakes, which were selected to the "rejected take-off" setting.

The tower contacted the flight crew, informed them that winds were 270 degrees at 27 knots (31 mph), and cleared them for take-off. The controller's wind report surprised the flight crew because it was higher than the 11 knot winds reported in ATIS Sierra. The captain recalls telling the first officer something like, "Roger, crosswind." The first officer recalls the captain saying, "Winds are 270 at 27. Are you ready?"

The captain began a reduced power take-off. He first pushed the thrust levers up and then increased power. He noticed a difference in the thrust being generated by the two engines, but the two engines matched as he increased to 90 percent. After verifying this, he pressed the TOGA (flight deck intercom) button and called out, "check power." The first officer responded that thrust was set at 90.9 percent. The captain applied a left control wheel correction, applied forward pressure to the yoke, and used variable right rudder to keep the airplane aligned with the runway center line. It felt at first like a normal crosswind take-off.

As the airplane was getting up to speed it suddenly swerved to the left, as if hit by a gust of wind, or as if the tires had hit a patch of ice and lost traction. The captain used full right rudder, but saw the airplane continue to veer left. As airspeed was increasing from 87 to 90 knots (103 mph), the first officer looked up and saw the airplane drifting left of the runway center line. He thought the captain was correcting back to the right, but the airplane suddenly yawed 30 to 45 degrees to the left. It appeared to the first officer that there was no directional control. He felt the rudder pedals with his feet and he believed the captain was applying full right rudder.

The captain saw the edge lights on the left side of the runway. He believed the airplane was going to exit the left side of the runway and, as a last resort, he reached down with his left hand and grabbed the tiller for a second or two. He attempted to steer the airplane back onto the runway using the tiller, but this did not work so he put his left hand back on the yoke. These decisions were quick. It was one of those moments in life when time feels as if it is moving in slow motion, but in actuality is moving lightning fast.

The captain used the right control wheel to keep the wings level as the airplane departed the left side of the runway. As he did this, he thought the ground next to the runway sloped down. He feared the aft end of the fuselage would slide down that incline and cause the airplane to tumble on its side. After the airplane had thoroughly exited the runway, the captain said, "reject" and tried to deploy the thrust reversers. He was unable to deploy the reversers because the ride was incredibly rough. The plane was shaking. Cabin lights were flickering. The passengers were jolted and surprised by the drama transpiring.

The airplane, crew, and passengers suffered two violent impacts before the plane came to a stop. It was totally dark in the cockpit. Neither the captain nor the first officer heard any engine sounds. Both were stunned and injured. They felt incapable of doing anything for what seemed like one or two minutes. They did not order the standard evacuation, nor did they begin the evacuation checklist. They had trained in these procedures and drilled them many times. They were unable to perform them at that moment. The passengers were momentarily leaderless.

Both pilots recall that the flight deck was pitch-dark. The first officer said that he could hear things going on in the cabin. He thought that he needed to make a public announcement, but did not. His next thoughts were about getting both himself and the captain out of the aircraft. As soon as he recovered from the initial shock of the crash, the first officer opened a cockpit window to his right and threw out an escape rope. He saw fire along the right side of the airplane. He got out of his seat and exited through the cabin instead.

Evacuation

Meanwhile, back in the passenger cabin, the damage inside the aircraft was considerable. A fire was quickly consuming the right side of the aircraft, worsening the situation. Upon looking out the windows and identifying imminent danger, the flight attendants commenced an evacuation, and ushered the passengers out of the burning plane with the help of two deadheading pilots. ("Deadheading" is a term in the aviation industry which refers to airline flight staff who are being transported free of charge and are not working on the flight.) The deadheading pilots did not go out the right side of the aircraft because of the fire. The cockpit door remained closed for the entire evacuation. By the time the first officer opened the door, everyone had exited the aircraft except the deadheading crew and one of the flight attendants.

As the first officer stood up, a deadheading crew member knocked on the cockpit door. The first officer opened it. About this time, the captain was trying to get out of his seat as well, but a dislodged flight crew bag was blocking his path. The first officer moved the bag, and with the assistance of the deadheading crew member helped the captain out of the cockpit. The deadheading crew member told the first officer that all of the passengers had been evacuated, and then the captain, first officer, and deadheading crew member exited the airplane via the door slide.

The three flight attendants experienced no problems with the escape slides, and the emergency exit lights were brightly illuminated, which proved quite helpful in guiding the passengers out of the aircraft, given the dark conditions. All passengers and crew exited the airplane via the left side doors and overwing exit. The flight attendants reported that the passenger who opened the overwing exit did so quickly and easily, in part due to the detailed safety briefing given to each passenger in the emergency exit row seats prior to every take-off. A bottleneck of people developed by the left overwing exit, and the deadheading captain directed passengers out via the doors or floor exits.

The cabin was illuminated because of the fire. The entire wing and wing root were on fire, which was most noticeable at the overwing exit. The deadhead first officer unbuckled his seat belt, turned left, and saw that the male passenger sitting in the exit row had got the door open quickly. There was a tremendous confluence of passengers trying to exit through the overwing exit. Five people were trying to get out first; no one wanted to be second. The windows were melting and popping. Passengers were screaming, "we're gonna burn" and "it's gonna explode." It was incredibly frightening for the passengers. Many people were trying to get out at the same time; some passengers were climbing over the seats to exit. There was so much panic that most instructions fell on deaf ears.

The other deadheading crewmember, a first officer, assisted several elderly ladies in evacuating, and re-entered the aircraft. He noticed the two pilots emerging from the flight deck being helped by the deadheading pilot, both obviously in pain. He helped them to exit through the door, and then came back for the first class flight attendant to help her, since she had an injured ankle.

The deadheading first officer assisted the injured flight attendant off the plane, and returned into the plane. He found the deadheading captain and the male aft galley flight attendant in the aisle as the plane started to fill with smoke. They met in the middle, over the wing, and began to check for anyone else who might have been left on the airplane. There had been a large number of small children on the flight. The two wanted to make sure everyone had evacuated. The aft flight attendant said it was all clear in the back. The deadheading captain made sure by asking if everyone had departed; the deadheading first officer said yes, and the captain told him to go back and check one more time.

By this time, fire and smoke were starting to come up through the floor. The crew were concerned the center fuel tank might explode. By the time they left the aircraft, the windows were starting to melt, and they were fearful there would soon be a breach; everyone knew it was time to get out of the aircraft.

Outside the aircraft, the first class flight attendant was in a great deal of pain and could not stand up. The deadheading first officer picked her up and carried her away, as the fire was growing fast. The center tank gave way, and a river of fuel ran north–south toward the nose, and fire was following behind it. It was starting to light up the entire cabin.

Luckily, the slides deployed properly. Since the landing gear was sheared off, the doorsills were not far above ground level. The evacuation slide was more a padded walkway. However, the gray colored slides might have been easier to see if they had been bright yellow or fluorescent orange.

Runway Conditions at the time of the Event

Inspection of the runway following the accident revealed that it was bare, dry, and free of debris. The first tire marks were found about 1,900 feet from the runway threshold. The aircraft exited the runway about 2,650 feet from the runway threshold, and continued across a snow covered grassy drainage basin area. It then crossed a taxiway and a service road before coming to rest about 2,300 feet from the point at which it departed the runway. On the night of the accident, the majority of air traffic was arriving at the airport from the south, and departing from the airport to the north.

Weather Conditions at the time of the Event

Both pilots were aware of the crosswind conditions, having been advised by Air Traffic Control that winds were 270 degrees at 27 knots (31 mph) before take-off. The weather observation in effect for Denver International Airport nearest the time of the accident was reported to be winds at 290 degrees and 24 knots (27 mph), with gusts to 32 knots (36 mph), visibility of 10 miles, a few clouds at 4,000 feet, and scattered clouds at 10,000 feet. The temperature was reported as 24 degrees Fahrenheit. Wind data had been obtained from the airport's low level wind shear alert system, consisting of 32 sensors located around the field which recorded wind speed and direction every 10 seconds. This information would have been used to determine a better estimate of the actual crosswind component at the time of the accident.

Leadership

The captain was interviewed four days after the accident while still hospitalized with injuries that included spinal fractures. He stated that he "was either knocked out or dazed" immediately after the crash, and did not recall how he got out of the airplane. Both pilots recalled that the flight deck was totally dark. The first officer said that he "could hear things going on in the cabin," and he thought that he "needed to make a public announcement" but he did not. His next thoughts, he said, were about getting himself, and the captain, out of the aircraft. The cockpit door was closed for the entire evacuation. By the time the deadheading first officer opened the door, everyone had vacated the aircraft except the deadheading crew and a flight attendant. Meanwhile, back in the passenger cabin, the damage inside the aircraft was considerable, and a fire that was quickly consuming the right side of the aircraft was worsening the situation.

As soon as the aircraft came to a stop, the deadheading captain unbuckled his seat belt, and, although injured, assisted with the evacuation that was already underway. He gave a description of the conditions inside the cabin. He said the panels in the middle of the row "had swung down and were still swinging." He tried to keep them out of the way as people went by, because he knew "the passengers could be injured because they swung so fast." Although passengers were bumping him as they passed by in the aisle, he finally got the panels up and locked into place. He then went toward the back and got three more panels locked up.

He said they hinged on the aircraft right in the aisle, and they were down and swinging back and forth. He pushed them back in place, and that was how he was hit. He said he was 6′ 3″, and was holding them back. Passengers holding babies hit him a couple of times. He jumped to the other side, and pushed a panel up, got hit by another passenger, and fortunately locked the panel back into place. He said the panels did not malfunction, they had just come undone.

As the fire grew more intense, the deadheading captain could see a breach in the cabin just aft of the exit row. He saw the emergency lights were on, but could not see past the breach because it was dark in the back. He saw the flames on the outside of the aircraft through the windows to the overwing exit. He said he did not feel any heat initially. The deadheading captain then looked toward the cockpit. He saw "the forward flight attendant was standing on one leg, holding herself up. The cockpit door remained closed." The forward flight attendant assisted with the evacuation of all the passengers. Then the deadheading first officer opened the cockpit door. The captain "was out of his seat between the pedestal and the cockpit door, and was in excruciating pain." The deadheading captain said he could see that "both pilots were seriously injured," and that "they looked dazed from the impact." The two deadheading pilots got the flight deck crew out of the cockpit and helped them through the door. The fire was on the right side of the aircraft. All on board evacuated through the doors on the left side of the aircraft.

Crew Mindset and Context

The accident flight was the first flight on the fourth day of a four-day trip for the crew—comprised of the captain, the first officer, and three flight attendants. The captain, the pilot flying, had accumulated a total of about 13,000 hours, with about 5,000 in the 737. The first officer had flown about 7,500 hours in his career, with about 1,500 hours in the 737. The first officer was the pilot monitoring. All appeared normal until the aircraft began to deviate from the runway center line. The captain noted that the airplane suddenly diverged to the left, and attempts to correct the deviation with the rudder were unsuccessful. He briefly attempted to return the aircraft to the center line by using the tiller to manipulate the steering of the nose gear, but was unable to keep the aircraft on the runway. Additionally, there were two of this specific airline's pilots who were on board as passengers while deadheading back to their base—the deadheading captain and the deadheading first officer. The two had flown the aircraft into Denver on the previous flight, and were passengers on the accident flight. Neither pilot cited any anomalies on the inbound flight, and they reported that all the aircraft's systems had operated normally.

Data Recorders

Bumping and rattling sounds audible on the Cockpit Voice Recorder have been time-correlated with the Flight Data Recorder, and were found to have occurred as the airplane exited the runway and traveled through the grassy areas adjacent to the runway. The aircraft reached a maximum speed of 119 knots (137 mph). It was traveling at 89 knots (102 mph) when the Cockpit Voice Recorder and Flight Data Recorder stopped recording. Physical inspection of the engines and information from the Flight Data Recorder did not indicate any evidence of pre-impact malfunction in either engine. The Flight Data Recorder data indicated that the number one engine power was reduced before that of the number two engine during the accident sequence, and examination of the engine indicated this reduction to be consistent with snow and earth ingestion as the airplane departed the runway. The Flight Data Recorder data also showed that both engines had been commanded into reverse thrust following rejection of the take-off by the flight crew, which occurred after the aircraft had already left the runway.

Landing Gear and Controls

A preliminary examination of the rudder system revealed no abnormalities or malfunctions. The main landing gear and brakes, which had separated from the aircraft during the accident, were found to be in working condition, as there were no signs of the hydraulic systems leaking or flat spots on the tires. The flight deck controls and corresponding control surfaces were found to be in the take-off configuration.

Aircraft Cabin

There was no obvious damage to the passenger seats, which were found secured to their tracks. The safety belts were all intact, although some showed evidence of fire damage.

Injuries

Of the 110 passengers and the 5 crew on board, some 38 sustained injuries, including broken bones, though everyone on board survived. Two injuries were critical, though both passengers' medical conditions were upgraded that evening. By the following morning, fewer than seven people remained hospitalized.

The captain was one of the critically injured. He was hospitalized with serious back injuries and bone fractures.

Investigation

Despite conflicting initial reports, and reported media inaccuracies, this accident is considered a "successful accident" due to several contributing factors: the flight crew's immediate actions, the presence of the deadheading airline employees on their way home, the fire rescue training that was developed and practiced for just such an event, and divine intervention or luck.

Incident response timeline:

1. Crash event happens at 18:17.
2. Crash reported at 18:18 local time by the air traffic control tower.
3. Initial dispatch—inaccurate location and limited information.
4. Denver Fire Department was on the scene of the actual accident at 18:23 to extinguish the fire.
5. Denver Fire Department confirms exterior fire out at 18:25; moves to fighting interior fire.
6. Denver Fire Department reports all fire out at 18:28 and initial primary search results are negative.

The aircraft's black boxes were recovered from the wreckage in usable condition. The Cockpit Voice Recorder did not reveal any apparent problem until 41 seconds after the aircraft's brakes were released, just before take-off. At that point, a bumping or rattling sound was heard, and the crew aborted the take-off four seconds later. Both recorders stopped working 6 seconds after that (before the plane came to a stop). At one point during the sequence, the plane's speed reached 119 knots (137 mph). Both the captain and first officer had outstanding safety records when the crash occurred, and were experienced pilots. Wheel marks left on the ground as well as initial reports from passengers and firefighters indicate that the plane was airborne, briefly. It is unclear at which point during the sequence the fire started. There was no snow or ice on the runway; however, there were 31 knot (36 mph) crosswinds at the time.

The flight crew that flew the aircraft to Denver before the incident flight were also on board, though not on duty, and reported having no difficulties with the plane during their previous flight. The aircraft was clearly in operational order before the runway excursion event happened.

Initial reports indicated that the plane could have suffered a landing gear malfunction that might have resulted in a wheel lock-up during the take-off roll, leading to the runway excursion. National Transportation Safety Board officials said that when the take-off began, the aircraft's engines appeared to be functioning properly, its tires were inflated, and the brakes did not appear as if they had failed or otherwise malfunctioned. They concluded that the landing gear had not caused any problems.

On July 17, 2009, it was announced that focus had shifted to a possible large gust of wind or a patch of ice. The pilot of the aircraft stated that: "My speculation is that we either got a big, nasty gust of wind or that, with the controls we had in, we hit some ice." The winds were reported at about 24 knots (28 mph) to 26 knots (31 mph) from the north-west, with gusts up to nearly 32 knots (37 mph), just before the airliner began its take-off roll northward down a north–south runway. The 737 is specified by design for a crosswind limitation during take-off of 33 knots (38 mph) on a dry runway.

According to an article published on July 18, 2009, in the *Denver Post*, flight 1404 had been equipped with "winglets"—curved, upswept structures added to the tips of wings—in November 2008. The article states: "The B-737 flight manual 'limitations section' showed a crosswind limit of 33 knots for a dry runway. But NTSB added that the manufacturer and installer of winglets that were on the airplane had published a maximum demonstrated crosswind component of 22 knots (35 mph) for winglet-equipped B-737-500s."

On July 13, 2010, the National Transportation Safety Board published that the probable cause of this accident was the captain's cessation of right rudder input, which was needed to maintain directional control of the airplane, about 4 seconds before the excursion, when the airplane encountered a strong and gusty crosswind that exceeded the captain's training and experience. Contributing to the accident was an air traffic control system that did not require or facilitate the dissemination of key available wind information to air traffic controllers and pilots, and inadequate crosswind training in the airline industry. This example illustrates surprisingly basic deficiencies in individuals' skills and in organizational capacity that pilots and air traffic controllers were unaware of as recently as 2008.

Interestingly, had this rejected landing event happened 100 yards further down the runway, the event would have happened directly in front of one of the airport's firehouses. Even though the event did happen close to a firehouse, the fire rescue teams were delayed by mistakenly being told when dispatched to head out of the firehouse and turn to the left. Had the fire rescue teams turned right instead of left, they would have been at the event nearly immediately.

When the fire rescue teams rolled up on the scene of this runway reject event, the firefighters doing the initial assessment of the accident noticed the pilot in the flight deck of the plane waving his arms towards the responding rescue crew members. The firefighters assumed that the waving arms were a signal that the pilot and first officer were OK—there were no problems in the flight deck of the plane. In fact, the pilot was attempting to signal to the rescue team that they needed assistance, and that there was an identified problem at the front of the plane. Is there a difference between waving for and waving off assistance during an event?

DISCUSSION

Let us identify some of the conditions that, although normal, combined to create this DFEO:

- Weather
- Skills of crew
- Warning and notification systems
- Quality of operational information
- Surprising outcomes happening simultaneously
- Design features of the wings—which create clear fuel efficiency
- Emergency response systems
- Operational leadership
- Normal system responses up to the point of the failure

This event is "classic" in the unexpected combination of relatively normal aviation performance variability, which produced a clear, and potentially life-threatening, deviation from an expected outcome—in this case a safe and on-time departure from the Denver International Airport. All of the conditions listed are extremely beneficial in creating the context for this failure. Keep in mind that this by no means is a complete list. The problem is that all of these conditions are also incredibly normal.

4
Workers Don't Cause Failure, Workers Trigger Failure

"Cause is not found in the rubble. Cause is created in the minds of the investigators…"

Sidney Dekker, *The Field Guide to Understanding Human Error*

QUESTIONS ABOUT YOUR ORGANIZATION:

- How long would it take to find a worker at my organization not following a procedure?
- Does not following an organizational procedure always create an immediate failure?
- Why is it when we have an event that we attribute that event to a failure by the worker to follow procedures?

REMEMBER: THREE PARTS OF A FAILURE

Every event you will have or have had in your organization is actually made up of three distinctly different parts:

1. **The Context of the Failure**: This is everything that led up to the actual event. This is the worker, the worker's mindset, the environment, the people, the hazards, the planning, and everything else that make up the story of the failure. Remember, in the context section of the event there is much information that the worker does not or cannot know—for example, how the event will end.
2. **The Consequence of the Failure**: This is what happened, the effect of the failure, the damage or harm, the deviation from the expected behavior. This is the ending of the event.
3. **The Retrospective Understanding of the Failure**: This is how the organization views the failure or deviation from the expected behavior. It is at this point that all knowledge is available about this failure. The retrospective has almost superhuman knowledge of every facet of this event. This is when most organizations look for a cause—a specific

moment in time when something bad happened—and use that cause to understand why the failure transpired.

CAUSE IS A CONSTRUCT—CAUSE IS MADE UP

How crazy is this going to be? I am about to commit safety blasphemy. I am going to tell you that workers don't cause failure. It is true, workers don't cause failure; what workers do is trigger a whole lot of weaknesses that exist in environments, processes, systems, job sites, and in the work or organization itself. These weaknesses are not new or unusual. These weaknesses are always present in your organization.

That is right—workers don't cause events. Seriously. But the problem with this statement is not what you think—it is not the reduction of accountability for failure from the workers. We are not trying to create an organization that gives all workers a free pass. Workers are always accountable for their actions—every single action. Just as companies and organizations are also always accountable for their systems and processes that bind and drive worker actions. The problem is the use of the word "cause."

Here is the problem. Culturally, we have become trapped by the idea that everything must have a cause. Most organizations use the term "root causes" when they do failure investigations. We call them "causal factors" when we thoroughly explore failure. We think if we can just find the cause of the event, and fix the cause of the event, our workplaces will be safer and our problems will all go away.

In the Freon case study at the start of this book, what is the cause of that event?

Could the cause be a stupid worker? Could the cause be a lack of supervision 8 hours a day, 5 days a week? Could the cause be the presence of Propane tanks in the Freon storage area? You read the case study. What was your idea of the cause for that event? Is there just one cause? Are there many causes? Is there a root cause? Is the cause in the Freon case study even necessary in understanding that failure? What improvement actions can we apply, based upon what we understand from this search for cause?

Organizations need cause. They must have something or someone to blame for such a stupid act. It's as if you cannot operate your facilities if you don't know what caused the most recent failure. Why is there such a strong pull to find a cause and fix the problem? Let's explore this idea.

Why do organizations have this almost zealous need to find the cause of a failure? The following reasons, listed in order of perceived operational need, may help answer this question.

You must have a cause for every event that happens at your facility because:

1. **Not knowing what caused an accident is downright scary.** It is simply not acceptable not to know how something happened. You can't tell your management, "we stopped being 'lucky' and, therefore, had a failure." You would go crazy and you would get fired from your job.

2. **Headquarters, management, or the regulator require a cause statement.** The regulators at my facility want to know three things immediately after a failure happens. Who did it? How did we let them do it? And where are we going to transfer that person, so it never happens again? If those questions are what the regulator wants to know, it is no surprise that the answers to those three questions are what we look for when we investigate and write up a failure.

3. **We have a need to punish the guilty.** You could also say that we have a strong need to tell the rest of the world how, had it been me doing that work, "I would not have done the same things the worker did in order to cause the failure." A better worker would have been smarter, or cared more, or not been so stupid. This need to punish the guilty seems to be almost instinctual in human beings. We probably don't want to punish the guilty; actually we want to assure that the failure that just happened cannot happen again. The only way we can do that to our satisfaction is to attribute the failure to the guilty person, and not to luck or to the system. We humans are a complex bunch when it comes to assuring stability in our lives and systems.

4. **We must develop corrective actions.** We must do something as soon as possible to bring our systems back in order. One of the strongest drivers to find the bad worker and identify a root cause is the fact that it is much easier and cheaper to fix *one* worker and fix *one* cause than it is to look at an entire set of processes and relationships that exist throughout an entire organization. Be careful not to fall for the easy answer—it has the potential to create a less reliable and more dangerous system, not performance success.

5. **We want to know how to adjust our organization's systems and behaviors to avoid the same kind of trouble happening again.** Finally, here is a reason I can stand behind. This is exactly how you should always analyze every type and every level of failure. Ultimately, you should only try to understand failure for what you can learn, not for who you can blame. It's too bad that this is the fifth entry on our list and not the first. Too bad that we must fight past the "eureka moment" of the human failure in order to get to the deeper issues, and discover the fundamental system weaknesses and conflicts. In many ways we construct cause in order to make our need for the creation of improvement actions quick, simple, and submitable.

Therein lies your challenge. All of the factors listed above for the Freon event (plus many more) caused the event to happen, and, in a way, all of those listed factors did *not* cause the Freon event to happen. Failure is the sum total of all of the things that exist in a specific worksite, and all of their relationships with each other. None of the individual parts of the Freon failure were sufficient to cause the overall failure—but all of these individual parts were collectively sufficient to cause the failure.

In a classic root cause analysis (RCA), your job is to deconstruct the event down to its minutest parts, analyze those parts, and fix whatever is broken. In Human Performance, you do almost the opposite. Instead of deconstructing the event, you construct the event context, and look not at the individual pieces of the event but at the relationships between those pieces.

I am getting a bit ahead of myself.

In order to build my case against cause, I have to have a small discussion around the way we solve problems. Before you jump to discredit this discussion, please allow a little conversation about how "cause" forces an organization down one specific path (mechanistic), while looking beyond cause allows our organizations to go down another, different path (holistic).

You must also talk about and introduce the concept of human error. We have discussed error all the way through this book, and we have discussed it on purpose. In order to have even half a chance of moving your organization away from the classic cause and effect trap that we all find ourselves in, we must give managers and leaders in the organization a logical reason to question cause. Knowing that errors happen even when a worker does not mean for an error to happen allows you to have a discussion about why human error is neither a cause nor an effect. More importantly, there is no relationship between error and effects. Error just happens; it either does or does not have an effect. Cause, according to the rules of logic, always creates effect. Error is not a cause, nor is it an effect. Error is simply error—an unintentional deviation from an expected behavior or outcome (DFEO).

MECHANICAL FAILURE VERSUS HUMAN FAILURE

Causal analysis is the best tool in the entire world to find and understand mechanical failure. There simply is nothing better. Cause also helps describe the laws of physics incredibly well. So let's begin this argument by saying straight out, "if you have a machine that breaks down, and you want to know both why it broke and how to fix it, use all the classic causal analysis tools in your toolbox." Those causal analysis tools will work perfectly, and you will get the exact result you want and need.

If my car breaks down, I want my mechanic to use causal analysis to determine what is wrong with the car. Even better, I want my mechanic to identify the "root cause" of the car's failure, then either fix or replace the part in question, and get my car up and running, in "factory floor" condition.

However, if your organization has a failure that includes people and technology, or people and automation, or people and machines, causal analysis tools are not the best choice in understanding "how" the failure happened. As a matter of fact, for those events involving relationships between people and machines, or people and technology, causal analysis is the wrong tool.

I am sure I don't want my doctor to use the classic root cause analysis procedure to fix me if I am broken somehow (nor would you want your doctor to do the same with you if you were broken). If I get sick, I want a doctor who thinks of everything that could be going on with my health, then slowly and surely examines me as a complete system. I certainly don't want my doctor to take me apart and look for what is wrong with me. If that were to happen, I would be more concerned with the doctor putting me back together.

Instead, the doctor tries to construct their diagnosis based upon a series of small (or sometimes large) symptoms that I may be displaying. Some of the symptoms I will tell the doctor I have, other symptoms the doctor will discover during closer examination and tests. All of these individual symptoms will now be analyzed holistically to lead the doctor to a complete way to diagnose what needs to be done to make me better.

The safety world has surrendered a lot of its ability to understand failure to a classic, cause and effect understanding of the world. That has had negative consequences for our organizations and for our job sites. It is a terrible thing that we don't look for and construct the whole story. Instead, we search until we find the first thing broken—often called the "eureka component." We call that first broken thing out and fix it.

Here's the challenge: that classic world view is not complete enough, so we need to make it big enough to capture all of the failure. Cause and effect implies a remarkably linear failure path, a mechanical failure. Human failure is never linear. Human failure always involves complex relationships: relationships between people and processes, people and technology, people and machines, people and other people.

One of the best places to show the importance of looking at the event as a system warning is found in the way we think of jobs for workers in the field. There is a vast difference between the work we think we give workers and the work the workers actually discover in the field.

THE DIFFERENCE BETWEEN WORK AS PLANNED AND WORK AS DISCOVERED

Your organization plans work. You scope contracts. You manage available resources. You have your daily planner, list, and calendars; you plan the crap out of every possible thing you do. It is a smart way to do business, and a powerful way to ensure mission success and reduce operational surprises. The only problem with having a carefully developed plan happens when the human worker actually starts following the plan. We all know that a plan is perfect until we begin to use the plan.

The following graphic helps guide this discussion. The solid line—the work as imagined or work as planned line—represents work performed the way everyone in the organization believes work should happen. Think of the work as imagined line as perfect work, perfect customers, perfect workers, perfect weather, perfect tools, and perfect processes. Everything that can go well is going to go well, and usually does.

If the solid line represents a perfect world, the dotted line represents reality. In fact, the dotted line seems to be much more representative of how work truly is to the worker. Our workers discover work; the workers almost never are presented with the work they are told they will get. And in order to be successful, workers must learn to adapt your organization's processes and procedures to match the work in front of them. Not vice versa. Workers can't change the work; therefore, they must change the processes by which the work will be accomplished.

The solid line—doing only what has been planned to happen—does not lead to mission success. The dotted line—adaptive behavior exercised in the field—is *always* how work gets accomplished. In fact, the dotted line is the only way workers ever create mission success. If your workers did not adapt, your important work would not get done. Given these ideas, it is vital to emphasize that the dotted line is not a terrible thing. You should not approach the dotted line as something you need to fix, or align more closely with the solid line. The

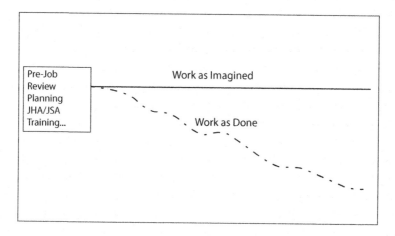

Figure 4.1 Adaptive Nature of Work as Performed

dotted line, work as reality, is always a part of the work. This two-line model precisely represents every job you have ever done. Since we know this model is always present when workers do work, we should use this model as a tool for performance improvement.

The space between planned work and performed work is the operational gap. In that operational gap lives a vast amount of information. This is where you learn about safety, as safety exists in your operations. This is very different from observing worker behavior or auditing procedural use and adherence. This is real post-job information about what happened when work was being done. Finding and understanding the difference between "work as imagined" and "work as actually done" is like finding the place where all your safety data "hangs out." This is a treasure trove of information. Seek understanding here in order to know how to better plan work the next time you perform this task, or tasks like this task.

We *plan* for safety in the pre-job activity. We do hazard identification work, job hazard analysis (JHA), job safety analysis (JSA), training and qualification— all for the job that we imagine the workers will do. We *learn* about safety in the post-job review and operational learning. Both are essential, but one cannot be successful without the other. For too long we have placed our safety emphasis on better planning. We now know that real operational learning comes from understanding the difference between work as planned and work as done.

The problem with all this operational "play" in the work our workers do is the lack of knowledge your organization has about the work environment. As workers adapt and improvise solutions to "discovered work," new dangers are also being discovered. In the environment of discovery, our workers are now dealing with hazards that we have not planned for, mitigated for, protected for.

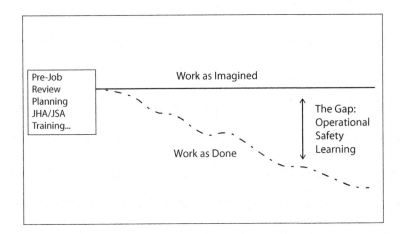

Figure 4.2 Operational Learning Happens Every Time a Task is Performed

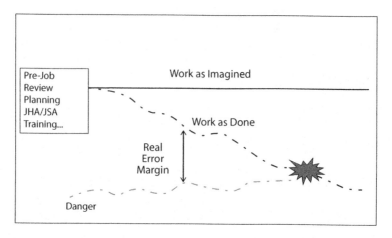

Figure 4.3   Actual Safety Margin

The best way to discover this information is to ask for it. You can do post-job reviews of successful work. Indeed, you *should* do post-job reviews of successful work. The only problem is that there is not much incentive to review successful work. We almost always are starting the next job before we complete the last job. That is simply our business reality. However, asking workers four quick questions which take little time or effort, and little to no money, will start to give you operational information, and reinforce a culture of "knowing" within your organization.

Tool: Successful Wrap Up

Four Post-Job Questions:

1. What happened the way you thought it would happen?
2. What surprised you?
3. What hazards did we identify and what hazards did we miss?
4. Where did you have to "make do," improvise or adapt?

Preprint these four questions on cards, the bottom of your job tickets, or as a part of a producton or manufacturing traveler ticket, and ask workers to spend a few seconds answering these questions. Pick these cards up. You are in charge of collecting and analyzing this information. Remember that once you have asked workers for the information, you will then have to do something with that information. They will expect to see an outcome or result.

The good news about the four post-job questions tool is that you will learn much about how work is done in your facility. You will be surprised by the places in your processes and in your procedures where workers in the field, on their own, adaptively perform their work. The more you know about how work happens, the better prepared you will be to help workers create safety in practice.

Case Study
Seven Stitches, a Broken Finger, Cardinal Rules, and Termination

Sometimes all you need to do is ask the question, "how?"

NOTE

This case study is not only about the event, but also about the tremendous need to understand that sometimes safety culture change can take years, and sometimes safety culture change can happen quickly. In this case, the organizational shift in mindset happened swiftly and at just the right moment. This organization had just sent their senior management team to an 8 hour Human Performance training session. The managers loved, and understood, the class and the concepts, but after the excitement was over, the management was left wondering what they should do with this information. How would they apply the lessons?

It is not uncommon for people to think that these ideas make sense, but not see a way to apply these ideas to their organizations. It seems silly that the managers can't see how these ideas—the ideas of the new view of system performance—are mostly about how they choose to react to failure, but in the application of the ideas, blindness is often the case.

However, in this scenario, the safety people actually pushed the management team to ask, "how did this worker arrive in this position?" Nothing about this case made sense to the safety people. Pushing management to pause and take a deep breath before they pulled the termination trigger not only saved a worker his job, but also powerfully helped shift the culture of the organization. This outcome was a product of a courageous safety person explaining an event differently, more completely to his operational overlords

After this initial discussion, there was a substantial change in the organization. Suddenly, the managers stopped asking how to apply the new ideas they had heard in Human Performance class, and started applying them. Action in this case made all the difference. Your job is to determine whether the action in question was the event, or the shift in thinking—or both.

THE REPORT

Problem Statement: A senior maintenance technician received 7 stitches, and broke the tip of his right index finger.

The event transpired when a maintenance technician reached in and "jerked aggressively on the belt" of a moving conveyor lift. The technician had removed the guarding, and was able to grab the belt directly in front of the roller apparatus. Because of the technician's choice of hand placement, coupled with the fact that the belt was moving, the index finger of his right hand was yanked into the pulley apparatus, causing the break and the injury.

This event was investigated, and the root cause of the event was determined to be the worker showing poor judgment in reaching in to a moving machine: worker error.

Pursuant to the company's policy on cardinal rule violations (Rule 1: *Worker exhibiting undue or excessive risk*; Rule 2: *Working on an energized system without LOTO* [lock out, tag out]), it was recommended that termination proceedings should begin for this technician immediately.

This senior maintenance technician, who had years of practical experience and organizational history, was about to lose his job for replacing a conveyor lift belt that, as it turns out, had never been replaced before.

THE STORY

A conveyor lift is a stable performance tool in most production facilities. If this piece of equipment is actively scheduled and actively receives preventive maintenance, conveyors simply work—almost in spite of themselves. Conveyor lifts rarely break, and if maintained, can run for many years.

In this case, the conveyor lift moved glass bottles into a loader system hopper. This lift was old and finally starting to show some wear after many years of steady operation. Because this lift worked on elevating bottles and dumping them into a bottle hopper, this lift moved remarkably slowly—10' a minute—much slower than the average speed a human walks. Conveyor belts that go this slowly last a long time and have little wear.

After operating on a production line 24 hours a day for over 20 years, this specific lift was finally starting to show some wear. The wear showing on the lift had indicated that it was time to replace the actual conveyor belt on this piece of equipment. The decision was made by the maintenance and production managers to replace the conveyor belt.

This specific belt had never been replaced, ever, by this manufacturing facility. The operating manual for this lift was secured. A new belt was ordered from the vendor, in accordance with both the belt numbers in the operating manual and the belt numbers available from the vendor. It was not difficult to order a replacement belt. The exact replacement, according to the numbers in the operating manual, was ordered and delivered to the manufacturing facility.

The maintenance on this machine was scheduled to begin on a Friday by the maintenance planners. It was also determined that because this specific belt moved so slowly, the belt replacement task would be done without stopping the production on the line. Instead, the decision was made to position a worker in this area of the production line, who would hand-carry empty bottles over the heads of the maintenance crews replacing and aligning the conveyor lift.

The Friday crew got the belt replaced, but this task took longer than was scheduled. This task had never before been done for this machine. Since this was the first time this belt had been replaced, the maintenance crew was learning to do the work while actually doing the work. In short, the job took substantially longer than anyone had imagined or anticipated it would take to complete. To make this situation even more complicated, the Friday crew was rotating, and would not be back at the plant on Saturday to complete the task. There would have to be a crew "handoff" on this job.

When the Saturday maintenance crew came in for their shift, they inherited a work task that was nearly complete. The new conveyor belt had been installed, but not tuned or aligned. At this point, although the belt was installed, it was not able to be operated until it had been seated and adjusted to run. To all intents and purposes, this should have been an easy job for this new crew. The production facility had other belts throughout the facility, though none ran this slowly. However, the process of adjusting a conveyor belt was not normally seen as difficult.

Belts are normally replaced with the machine off, the machine guarding removed, and with the power locked and tagged out. However, belt adjustments can only be performed with the belt operating. Most conveyor lifts have belt adjustment bolts that can be turned slightly to fine-tune a belt so that it operates in a straight line and doesn't work its way off the belt rollers. This was the case with this particular conveyor lift. Machine guarding was replaced on the belt rollers, the power was unlocked, and the machine was turned on in order to adjust the belts.

During this time, as was the case during this entire maintenance activity, the glass bottles were continually being hand-carried and dropped into the bottle hopper. Production had not been slowed or hindered for this maintenance activity. It is also necessary to note at this time the fact that this production facility was under heavy security because of what the company produced. Virtually every square inch of this facility was constantly monitored by video recording surveillance. Every second of this event was recorded and could (and would) be analyzed for a clear cause.

The Saturday crew was made up of a senior maintenance technician and two other lower level maintenance technicians—a normal distribution of skills and talent. The senior maintenance technician in this case was well respected, with 20 plus years of experience in the facility, and was known as a safety leader. This crew was knowledgeable and capable, and normally needed very little supervision or direction.

The crew began the process of completing this conveyor belt replacement task. At around 9:00 am, the work began, and if everything had gone as planned it would have been done in a half hour or so. Two and a half hours later, the belt was not cooperating and was not yet installed.

When observing the video of this event, it is difficult not to notice the level of frustration of the workers increasing almost minute by minute. Nothing the senior technician did made any difference to the alignment of this belt. The video shows the technicians using a wrench to turn the set screws many times. Normally, these adjustments are small, but in this case, the crew could not get this belt to align correctly with both the end rollers and the tracking wheels. Both of these systems keep the belt reliably operating. During this time, the crew removed the machine guarding in order to assure the correct alignment of the belt.

At about the three-hour point in this process, the senior maintenance technician, moving up to the top of the belt—the section that was giving the crew the most trouble—grabbed the moving belt and jerked it with all of his strength. In the process of trying to force the belt into place, his hand was pulled between the belt and the belt rollers, breaking the tip of his finger, and lacerating his finger as his hand pinched out of the belt. No other worker, neither the other maintenance workers nor the worker tasked to manually carry bottles to the bottle hopper, attempted to stop this worker or even noticed his hand grab the belt.

THIS ACCIDENT WAS PREVENTABLE…OR SO THE ORGANIZATION WISHED…

The worker was injured. During the organization's normal, traditional event investigation, the decision was made that the cause of the event was the worker's poor choice in grabbing a moving belt running on an energized machine. If the worker had not grabbed this belt while the machine was running, this accident would not have occurred. If the guarding had been in place, it would have been impossible for the worker to stick his finger between the belt and the rollers, therefore this event would have never transpired.

In fact, both of those notions are true. Just as the following ideas are also true. If the belt had been the right size, the worker would not have become frustrated enough to try to force the belt into place, and this accident would not have happened. If production had been stopped, perhaps the crew would not have felt time and production pressure during this maintenance activity, and with the pressure relieved, the accident might not have happened. If the belt had moved faster than 10' per minute, the technician would never have grabbed the belt, because a fast moving belt is a dangerous belt, and therefore the accident would not have happened.

The idea that by identifying the things that the worker should have done in order to avoid this accident the organization would be able to either change history or prevent the next event is foolish, but extremely attractive when answering the question, "why did this happen?" It is also a form of malpractice—by fixing only the worker, the company is allowing work to take place in ways that actually remove the workers' ability to control their work environment. Remember: operational surprise removes worker decision, and creates worker reaction.

Let's look at the conditions over which the worker had little or no control. I have made a partial list of the conditions that existed in the making of this failure. Notice how little control of the actual work environment the worker had in the context of this failure.

Conditions beyond Worker's Control:

- Maintained production status at facility
- Belt moved exceedingly slowly, giving the impression of manageable risk
- Belt had never been replaced since installation of equipment
- Belt could not be adjusted using adjustment screws
- Belt design and tolerances had changed
- Belt part number had not changed
- New belt would never fit in current machine configuration
- Belts must be adjusted while running
- No communication between maintenance crews due to schedule turnover

WHY THIS STORY IS DIFFERENT

With the new found knowledge of "another way to look at this event," the safety crew at this production facility started to try to understand the event from a much different point of view. The facility safety people knew that as soon as word of this event got to the corporate level, their choices would be significantly reduced. If corporate found out that this worker had stuck his hand in a belt that was running there would be three questions: Why was this device not locked out? Where was the machine guarding? Why haven't we used our cardinal rules to fire this guy?

So, the safety people made certain they had the story of this event *before* they called corporate to make their notification. Their belief was that if they could tell the story differently, more completely, perhaps they could cause a change in the course of the future.

It worked. By reporting the whole story to the powers that make decisions, the safety people ensured that the company's actions changed from the traditional "blame and punish" to a more "diagnose and treat" model. No one got fired. No one had to justify breaking a cardinal rule. The event didn't change—the way the safety people talked about the event changed, and that created a different conversation with the corporation.

How did the response to this event follow a different path? In short, the production facility made a conscious decision to respond differently to this failure. Even though this failure contained all the corporate triggers to immediate discipline of a worker, the trigger did not get pulled. The safety staff at this production facility presented a more complete contextual understanding of the failure. In the long term, they made the facility and its employees safer.

DISCUSSION

Some questions you should consider about this Case Study:

- Is there a need for "cardinal rules" in any organization's safety program?
- Who gets to draw the line that says a rule has been broken? Every rule is built with contextual and political flexibility—rules are always open to some interpretation.
- Does past adherence to a rule absolutely dictate future use of the same rule?
- Are there cases when a cardinal rule has forced the organization to make a less than optimal decision?

5
Change is Better When You Manage Change—And Change Needs to be Managed

QUESTIONS ABOUT YOUR ORGANIZATION:

- What changes have been made in your organization recently that have been especially successful?
- What made those changes successful?
- What did your organization do to make the changes stick?
- What did your managers do?
- What lessons can you take from other changes within your organization that you can replicate in order to help you change to your safety program?

MANAGING THE SAFETY CHANGE

You are now asking your management team to go on a journey. This is not an easy jump; the jump to a fundamentally different view of performance reliability does not occur naturally. You will serve as guide, teacher, enforcer, and diplomat for the new view and new operating parameters—and this job will never go away. You will be the conscience of the organization. You will be the leader and mentor. You will also be the follower and student. Just remember: this journey is well worth the effort. You will never know how much good you are doing for your organization. Just know that you are doing the best work you have ever done.

The biggest enemy of safety change is dogma – the belief that we already know the answers to the questions our organizations are forced to ask. If we know what the problems are, or similarly if we believe we already know the solutions, there is no need to go in to the field to learn more. There is no need to change. If we already are convinced our organization is doing the right things, just not doing those right things good enough, there really is no need to change. That is why our organizations keep doing the same corrections over and over without a different outcome. In order to change our organizations must believe they do not have the answers to the safety questions. Our organization must know that they do not know.

Change does not just happen. Change is hard. Change must be managed. If you ask a group of people if they like change, some will say they do. If you ask that same group of people if they hate change, some others will say they do. If you ask a third question, "how many of you don't mind change, you just don't like it when change is forced upon you without explanation or input from you, or without regard to the potential consequences that the change will create?" all of the people in the group will agree to the truth of this.

THE JOURNEY YOU AND YOUR TEAM ARE ABOUT TO EMBARK UPON

We can't force change. We must manage change within the boundaries, culture, and context of our organization. You must think about how you want the change to happen, and then guide the process, all the while realizing that change management is never complete—the process never stops.

Here is a secret weapon for managing change. Use this weapon early because it has the potential to "color" the way you talk about change with the workers and managers of your organization. The secret is that all new knowledge demands building (or finding) a bridge to older knowledge in order to make the change. The best way to facilitate change is to attach new knowledge to old knowledge in a way that moves people to a different way of viewing the same old problems.

In other words, telling your managers that they have been managing their workers badly will never win any hearts or minds. However, starting the conversation by reinforcing the excellent work that has been done in the past around operational safety and then moving to ways we can "tweak" the current management style to represent the new view of safety performance will be much more successful. I have mentioned this previously, but it is worth stressing that your managers have managed safety to the very best of their abilities.

And that message also applies to you. Take the time to become educated and build depth in these areas. Don't be afraid to meet people who are on this same journey. Those people are a resource for you to use in managing your company's change. Read everything you can find on systems performance, error theory, complex adaptive systems, and Human Performance. Why should you read everything you can find? Depth matters, and depth of knowledge in this area can take a long time, but it is well worth the investment you will make. Depth of knowledge is not only beneficial, but also vital to being able to have the strength (both actually and emotionally) to manage this change every day for the next couple of years.

The other change secret I will share with you is not as pleasant. This new view of safety performance knowledge seems to need time and incubation before people in your organization will "get it." I learned very early, and sadly several times, that this discussion has to start at the beginning of the process. You must teach the fundamentals before you can talk about advanced concepts. Don't start with an extreme push for a "blame free" workplace before you take the first steps. Have the discussion about why turning first to worker behavior almost certainly ensures you won't identify the larger systemic issues on your job sites. Plus, once people get the impression that you are selling a "get out of jail free card" you will have a struggle to regain control of your message—and your organization.

Lesson number two is as profound: you don't have to take the resistance to these ideas to heart if someone says the ideas won't work—or worse yet won't allow the facility or organization to hold workers accountable. The idea that workers will "run wild and break all of the safety rules" is not an attack on you, but a desperate attempt to stay with what the person knows.

Remember, change—even for the better—can sometimes be a bumpy, but worthwhile road.

You have probably heard the phrase "change management" bounced around for a few years. It is simply the process of changing how your organization or corporate culture manages systems. It sounds scary, because human nature doesn't actually like to change. It's evolving from the "we have been doing it this way for 100 years" idea. Humans grab onto a set of ideas or values, and we become threatened when those are questioned. I think that calling it "evolve management" would make it easier. You are most certainly changing things, but these changes come from wisdom and building on the past. It is evolving. You are not throwing everything away. You are building on the past only to make it safer, communicate better, and create greater production. You are helping your organization evolve.

Your job will be to challenge a significant part of your organization's culture, its systems and values. It is the classic "changing the way we have always done it" argument. It sounds daunting. But it can be done. You will have completed the first part of your change management program already when you diagnosed that the safety system needed to change and evolve. How do you eat an elephant? One bite at a time.

Once you get the worker to buy into your mission then the changes will begin to happen immediately, but slowly. Keep the faith and keep your message.

Change is a lot like aging. It is not something that you may want, but it is happening never the less. Change is not evil or scary. Like aging, it can be a little difficult, but it is inevitable. This is why communication and diligence are so crucial in making successful change.

WHAT YOU CAN EXPECT

Managing change should be a planned and understood activity—a major part of your plan, not an afterthought or an accident. You should be acutely aware of what you want to do, where you want to start, and how you want to make change happen. Remember, you are in this for the long haul. Move deliberately, and with a clear goal in mind. Most importantly, get started. Nothing can happen until you start the change.

Different groups in your organization will need different strategies. Planning for your organization is clearly not "one size fits all." You will be asked several times in this book alone to identify groups or teams that are ready to make the change. These special groups will simply need a strong foundation of knowledge, a path forward, and your indirect support. The slow changers in your organization may need much attention. Think about how you will make your way into each of your potential worker audiences, and then customize your approach to meet these groups' needs. How long will this take?

Be ready for resistance. Prepare an answer for the question you will get a bunch of times while you are on this journey: "Is this another flavor of the month?" Know how to answer that question, because you will be asked. Human Performance isn't actually a program like other safety programs. There is no set recipe for how to start a program, and there is never an ending to the effort. You will never have a job site that is completely safe. Couple those complications with the notion that you will never be able to measure the events that didn't happen. Be creative and make your own milestones for both you and your workers along the way.

What does success look like? That is entirely up to you and your organization. Think about this question and have answers at several levels in your organization. Success for managers is going to look different from the successes of your workers. That is not only fine, but also predictable and beneficial. I am not sure that you will ever know when you have made it to the end of this journey. Workers change all the time, and your safety and reliability systems are always moving slowly towards failure. Given these two factors, it is probably safe to say you will never be done.

What you actually are doing during this change management activity is keeping the inevitable failures that already exist within your organization from being successful. Everything you do to increase your organization's ability to learn makes your organization safer.

WHY LITTLE CHANGES MAKE A DIFFERENCE TO YOUR ORGANIZATION

In my opinion, the most important and in many ways the easiest change you can make is to change your management team's reaction to failure. Work on this immediately. The pay-off for this type of change is enormous; the cost to make it is low. Let's face it, this change boils down to a collective choice to respond differently when a failure happens. You will have to gain commitment from your team. You will also have to make this change sustainable every chance you get, but it certainly is not too much of a stretch for managers to make. The trick is to be there to both remind them of the new view, and advise them what to do and how to proceed. Remember: this is a new way to respond to a rather old problem.

Change happens incrementally, one small step at a time, one person at a time. That is how you will have to think about this change management strategy—one worker at a time. What you are honestly doing is creating a whole lot of new conversations with your coworkers about these new ideas. Your job is to talk about Human Performance with everyone you can. The more you talk, the better the change will be in your organization.

WHY YOU SHOULD HELP OTHERS

The best experiences I have had during this journey to Human Performance— by far the most significant learning opportunities I have ever experienced— happened while I was helping my peers go on this same journey. For some reason, when you look from the outside in you have a much clearer vision of what is happening. My guess there reason that outside eyes see more has to do with knowing that you don't know. When the organization is not your organization, you get operatonally smarter in your own organization , and you leave with so many excellent ideas.

Working with other organizations allows you to reframe your issues and problems through the lens of another set of managers and workers. In essence, the safety performance observations you make are less personal and more systemic, because you don't know the people so you have to observe the system.

Case Study
Nine Senior Managers, a Million Different Opinions on How to Handle a Problem...and Nobody Willing to Change

MANAGER TYPES

Managers seem to be managers the world over. Guessing what type of attitudes are present in the management commitment meeting to shift the safety focus and event learning effort is not difficult. In this particular conference room, we had 9 managers. Each of those managers represented a different organization function for this organization. Each manager, although all employed by (or, in this case, owners of) this organization, had different incentives and understandings of what was necessary to the organization. These managers shared the common greater goal, but not necessarily the daily operational goals. That difference was noticeable in the approach these managers took to the new ideas.

During this meeting, it is safe to say that all types of resistance were exercised. There were unlimited reasons not to try managing safety with the new view ideas. It seemed during the first several hours of the meeting that all the classic pushback personalities were exhibited.

Here I will present you with the managers, and their reasons why this change could never work—and some suggested strategies to counter some of these arguments. Remember, these managers do have the best intentions at heart: their problem is around assuring stable performance in an unstable world. Try not to take any of these "pushbacks" personally. Listen to what they have to say and counter their comments with ideas that build bridges to Human Performance. Be passionate in your pushback, but don't be mean or judgmental.

The "I am too Busy" Manager

This manager has just returned from the UK and leaves for China in the morning. This manager is always busy, but the problem is made worse by the enabling behavior that all their underlings constantly discuss. Being afraid to ask this type of manager to block out some time is a mistake. Not asking for this manager's time is simply not respecting this manager enough to help them understand the extent of the problem. Don't fall for the "give it to me in a couple of hours—the 10,000' view—'cause I can get you that time." It is wrong to assume that the manager has had any exposure to Human Performance ideas—and, without a solid foundation, your program will fail.

Help these managers find open opportunities for conversations in their schedules. My favorite "too busy" manager finally gave up when I said, "let's meet Sunday afternoon at your country club, and have this conversation over a beer." It simply made it hard for him to tell me that he was too busy. Find ways to leverage time these managers think is more open and available than their normal schedule allows.

The "We're Different" Manager

This is the manager that wants you to know just how different the work, people, area, and style—and almost any other delineator you can name—are in their organization, and how this makes a program that looks at systems and organizational drivers a "non-starter." These managers need to have some time to process, and to apply the new ideas to their environments.

Of all of the management pushbacks that are to be encountered, this is the most normal and most predictable. Here is our secret: give these managers the satisfaction that their organization is different. Tell these managers that we will have to customize the ideas to fit their special world. Let them know that their organization is unique and that this will be a tailored effort. Then tell them that these ideas around human beings are universal and fundamental. The key here is to make this argument create less impact by agreeing that their world is unique and that their skills are astounding.

The "I Already Know this Stuff" Manager

These guys are tough. Sometimes these managers honestly do know this stuff and are your greatest asset. Other times these managers are working hard to maintain "credibility" among their peers. The best course of action here is to test their knowledge, gently, and see how much they actually do know. If it is "all good," then go with it. If there is no knowledge, play to their egos, and build a bridge to them that makes them feel included.

Refer back to these managers in your conversations and examples, and play up what is exceptional about the safety management philosophies that currently exist. Allow these managers to become experts. Use the "as you may know" type language in your discussion. Help these managers "fake it till they make it." Use their presumed expertise to your advantage. It works.

The "We've Got to do Something, Now" Manager

Also known as the "hair on fire" or "do anything—'cause it's better than nothing" manager. These managers have to be carefully and kindly coached, immediately, towards acting less emotionally and more deliberately. It is crucial to emphasize how much an attitude like this builds upon itself and the rest of the management team. Help these managers understand the process for a program like this, as well as the process for change that exists within companies. The key in this case is to help the managers realize the value of responding deliberately, calmly, on purpose, and with an understanding of the context.

Look for the most valuable things that you can do immediately to have the largest safety impact, or, as we call them, "the low hanging fruit." Build a sense of deliberate action into the safety management plan for this organization. Cheat your action plans towards the sooner side, as opposed to the later.

The "Flavor of the Month" Manager

This manager is always one conference workshop or airline magazine article away from the "next big thing." These managers honestly do mean well, and are excited about the "next big thing." It is not hard to motivate them toward change. But once it gets started, it is hard to keep this manager committed to the change. You know this type of leader, so therefore you know that what they want is an answer to their organization's troubles.

At first this manager will be your strongest ally. You must keep these managers interested and excited. This group is always ready to be a part of the change, and is usually able to help with the instruction, design, and implementation of your Human Performance program. Use them and let them shine. Then keep them excited with books and articles, and most importantly a part in the planning and discussion of what is going to happen next. You probably cannot over-communicate with these managers—so communicate with them often, and in pretty considerable detail.

The "I Have to Have a Way to Hold Workers Accountable" Manager

This manager is probably the most common manager you will encounter, even in your own management team. These aren't bad guys. They clearly want their workers to work safely and not to have accidents. The problem is that they see the world in the "crime and punishment" view of managing safety, and you want to move them more to the "diagnose and treat" model of safety management. Sometimes this is difficult, and sometimes you will never make any difference to these managers. The good news is that some of the "old school" managers will change, and become your strongest and best program supporters. When that happens, it is well worth the effort.

If you are faced with a room full of hard discipline managers, remember that these managers are often at a crossroads in their career. They must either change or get out of the way. Sometimes that discussion is exactly the discussion you need to have with these managers, and it is a hard discussion to have with them. It is also clear that many of their opinions about the old ways of managing are indicators of both some fear of the new and not being fully aware of how to make this change. When this crew doesn't answer your questions, it may not be them—it may be your questions. Be flexible and adaptable, and keep trying.

Build bridges to outcomes that are realistic for these guys. Tell them that you will help them better understand the work conditions and environments surrounding the workers, and that in doing so you will all see improvements in the safety outcomes. But also tell them that the outcomes are not the goal— the program is the goal. These guys will slip back to the old ways of thinking pretty quickly, and that is probably predictable, and OK. Be ready to coach these guys a lot—but coach them from a place of compassion and respect, not anger and impatience.

There are three things that are relevant to these "old school" managers:

1. **Your Credibility**: Make sure you know what you are doing and what you are talking about. You probably will not get a second chance with these guys. Be right and check your work.
2. **Your Attitude**: Don't be smug or overly theory based with these guys. Speak in real terms and give them real information. Don't ask this crowd to behave differently, tell them in clear and understandable terms what you want them to do differently, and why these different responses matter.
3. **Your Program Results**: These guys want to know what has changed, what has improved. You can't measure things that don't happen, but you can talk about changes in attitudes, communication, and early problem identification, and excellent catches that are happening in the field. Tell these guys stories about safety successes that are happening at their sites.

Here's the good news about this specific group: I find that this group of managers can become the biggest advocates for a Human Performance program. It takes a lot of work to turn this crowd, but the pay-off is often downright worthwhile. You can always ask them to try this new Human Performance stuff, and add that if they don't like it they can go back to their old ways.

The "I Want to do the Right Thing, but Don't Know What the Right Thing is" Manager

If a manager telegraphs this type of attitude to their leadership team, the message can be somewhat confusing, and almost dangerous. Why is it confusing and dangerous? Having a manager who admits publicly that they don't know what to do next can be a bit frightening for a team. Don't underestimate this manager type. These managers are in a position where they know that they have a problem, or potential problem. They are at pains to know what to do about this problem. This is a management team on which you can have a considerable effect. The problem here is pretty painless.

The solution is to give this management team an opportunity to learn, as a group, the fundamental ideas and concepts of Human Performance, with you as their guide and coach. Give this manager as much information about what other managers in similar conditions are doing as you feel necessary and appropriate. Just knowing that there are other managers who are learning and changing is often enough to reduce a manager's anxiety, and increase their potential and desire to learn new skills and try out new ideas.

The "I Have Done it This Way for 50 Years" Manager

These are the old "trapped in a rut" managers. This is a fascinating form of resistance to change. One of the most intriguing things about the idea of not changing something that has worked for the last 50 years is the fact that this method *has* worked for the last 50 years. Why change a process that seems to be working to a standard that the organization is willing to accept? Therein lies the problem—what the organization is willing to accept as their safety baseline threshold.

As the world, and more importantly the work that your organization does in the world, becomes more complex, the types of failures you experience in that world also increase in complexity. Suffice it to say that the old ways we managed phone numbers—memorization, phone books, important number cards—is not the way we manage phone numbers now: smartphones remember all of our numbers. It is the same for the way we manage all types of work, especially safety. We must change to keep up with the new complexities that we are constantly facing. Managers must have that knowledge—the knowledge that we must adopt new ways to view our workplace challenges.

The "I am Going to Agree with Everything You Say, but Not Change" Manager

This resistant manager type is the most difficult to change because this type is the hardest to predict. It is unclear whether this type of manager looks at you and nods to get you out of their office, or if they have the best intentions for change until the least amount of resistance appears, and then they retreat back to the old ways of doing business. What's worse, during the meetings and training sessions the managers usually respond as if the information has operational value, and presents ideas and strategies that these managers are willing to try. If your feedback systems are all telling you management is on board, it becomes extremely confusing when management class participation behavior does not track with management's safety decision making.

If management gets these ideas, or if management acts like they get these ideas, either way your organization's change is coupled to a manager who says the right things and then behaves in an entirely different way. It goes without saying that this type of manager must also exhibit this type of behavior in other areas of the business. In fact, this may be the beginning stages of understanding some systemic problems with the organization.

So what should you do? There are two suggestions that arise early:

1. If your managers are not able to answer some fundamental questions about their safety program, it may not be the managers in question. It just may be the questions being asked of the managers.
2. Look for ways to change the management safety system so that the safety system produces totally different outcomes from these managers. If you can't get the managers to move to the mountain, move the mountain to the managers.

Managers are people too. Ultimately, when we talk about the organization or the system we are talking about the people who make up our workplaces. It is essential that we understand that the same pressures we are observing in our workers are also having an influence on our leadership. Leadership is also a system outcome; however, leadership pressures may come from shareholders, regulators, or even the customers. The management behaviors we experience are an outcome, and we are best served by being aware of what is driving a strong need to punish, or pressures not to change systems. There are always answers to these questions. It is our job to help discover those answers.

Managers want the same outcomes that you want for your organization. That is vital to remember. Facing off with a manager in an argument between the new view and the old view may seem like a fight worth having, but remember that our job is to build an operational bridge from the old way you ran the business of safety to the new way you desire to manage safety. Big, loud arguments may seem important, but if a manager feels like they have lost a battle to protect the "tried and true" ways we manage safety, have you honestly won anything? You want your managers to be a part of this change—and share in the successes.

Probably the best advice that can be given when engaging management in the many discussions around changing the way they respond to Human Performance issues in your organization is to build relationships through conversations—not through browbeating and lecturing. Change happens through dialogue. A Human Performance program is a change in the way your organization manages safety. Move your managers by talking to them about what is admirable in the way they manage safety, and where there is potential to develop a new way to see work and workers.

DISCUSSION

Ask these questions about your organization:

- What type of managers am I dealing with in my organization?
- What individual strategies have I set up so these managers can be successful?
- What success stories can I tell this team to keep them both motivated for this change and encouraged that this change is having the desired effect?
- Have I assumed a level of knowledge and skill in safety leadership that is not there yet?
- Am I accessing reality? Am I working with the management team? Is it the case that I don't have the management team I wish I had?

6
Thinking about Where Failure Will Happen

QUESTIONS ABOUT YOUR ORGANIZATION:

- Where will the next event happen in your organization, and what do you think that event will be?
- Are you surprised when events happen?
- Are you ever really surprised at who has events at your facility?

TOOL: ONE CARD BIG ANSWERS

One morning, at either a normal safety meeting or some pre-job activity, give every worker on the team a 3"x5" card and a pencil. Ask them to take a second to write the answer to these two questions on the provided card:

1. On this job site, what is the next accident we will have?
2. Where will it happen?

Have them answer the questions quickly, and make sure that they don't overthink the questions. You actually want their gut level and instinctual reaction. Have them return the cards—you don't need names or worker identification information at all, just pick up the completed cards and see what happens.

In about 10 minutes' time, you will be given possibly the most valuable, most specific safety assessment information you have ever received. It is not an incredibly scientific method, but ours is not an incredibly scientific business. This method is a highly effective way to gather information about where these workers believe the highest potential for failure exists on their job site.

The only caution I have on this exercise, as with so many other tools we use, is not to overuse this method. This method is most effective if used sparingly, and less effective if used all the time. Look for ways to find variations on this theme to gather information from your workers. There are about as many ways to get this information as you have time to devise in your head and apply in the field. Creativity clearly counts for much in this area.

HUMANS SUCK AT PREDICTING THE FUTURE

You can't ever predict the next accident. Nobody can predict the exact failure that will happen next and its exact time. Too bad for us, because if that person existed he or she would be valuable—probably the most valuable person on the planet.

Human beings are bad at predicting the future. We do many things extraordinarily well, but prediction is not a strong skill.

Yet because human beings are quite good at assessing and understanding immediate contextual and conditional cues as they happen, we assume we are seeing the future. In fact what we are doing is simply applying what is happening now to what will happen next. Because there is an infinite amount of possibilities of what could happen, we have to have some method to limit those possibilities, and what we do is use what we know. Unfortunately for us, what we know now is not necessarily what will happen next.

Remember: failures tend to be unexpected combinations of normal performance variability. It is also worth reminding each other that the parts of an event are not as interesting as the relationship between those parts.

What we are not particularly adept at predicting is the space between people and systems. That inner space is marked by system complexity, which is the area where just knowing the people and the system is not sufficient enough to represent the universe of complex outcomes that could happen.

You can get pretty good at predicting which part of your organization is most prone to fail. It is possible to look at a task or a work environment and identify places within its processes where you don't have multiple layers of defenses, or places where there is immense pressure on production at the cost of making room for safety. It is even possible to identify places within processes where the interface between the worker and the technology the worker is using is not effective.

An important theme in the new view of safety is that the places where our workers think we are prone to failure are the places where normal performance variability may become amplified and breed failures. The power of not looking for the next accident, but, in fact, looking for the next place where a failure could happen is incredible.

So, if we have leveraged as much value as we possibly can from simply asking workers to care more about safety, what should we be doing to make our workplaces more defended and more able to fail safer?

WHY CRITICAL STEPS MATTER

> "The problem with uncertainty is that more bad things can happen than will happen." Plato

I am amazed every time I go into the field by the quite remarkable relationship between the number of things that go wrong and the number of things that *could* go wrong. Take any construction site you have ever seen in your entire life and think about all the multitudes of things that could go wrong. If you were purposely to have a bad failure, a construction site would probably be the best environment in which to have one. Accidents are really very rare, especially considering how much potential there exists in our organizations (and our world) for accidents to happen.

Everything that you could possibly need to have a failure is already present in your workplace. All of our workplaces are collections of latent processes, conditions, and components that are lying in wait for a worker to trigger failure from the midst of normal operations. These latent conditions, tasks, processes, people, pressures, and activities all work in combination to create a complex, adaptive organizational context—many things can (and normally do) work well together, or they can work together to fail.

You can't fix everything. You can't defend against every possible hazard at your worksites. If you tried to fix everything, you would go broke and crazy. So what should you do? One option is "dumb luck," where an organization hopes and prays that they can pick out the places of highest risk based entirely on luck of the draw. Surprisingly, lots of organizations do just that—operate by the belief that if the workers don't screw up, they will not have any problems.

There is another option to think about when thinking of how work happens in your facility. Remember our discussion about work as imagined or planned, and work as actually done? The space between the organization's idea of work and the workers' idea of work is where learning about safety happens. Finding the space for learning helps, but within that space is a way to prioritize individual tasks and activities, based upon risk and potential outcomes.

We call this process "the identification of critical steps." By identifying and understanding which parts of a process are most prone to have failure as a consequence, you can then layer those same critical steps with defenses against the predicted failure. The organization, with this knowledge and tactic, can then influence systems to create "survival space" around their workers.

IDENTIFYING CRITICAL STEPS

Let's define three terms before we go too far into this discussion:

1. **Normal work activities**: The universe of all actions, tasks, and activities that when combined make up the work being done.
2. **Safety important step**: A safety important step is a procedural step, process, or activity that exposes products, services, or assets to harm, potential harm, or an immediate degradation of defenses.

3. **Critical step**: A critical step (and there are many definitions for this term, depending on your industry and the work you do) is a part of the work process such that if this step failed it would have an immediate undesirable consequence for either safety, security, or environment, or politically—for the organization, the public, or workers.

Now that we know what we are talking about, let's discuss the immensely serious way to parse work in order to understand safety. To begin with, let's set a basic ground rule about why these steps are necessary. The most significant safety management tip in the world is pretty straightforward: never place a worker only one defense away from a failure. Really, the only way the organization can ensure more than one layer of defense around a worker when it matters is to know how to understand the importance of the work steps.

Who knows the critical steps of a process? The workers who perform the work are the experts on that system. The workers can tell you where the process is especially strong, and where the process depends more on the worker to create safety than on the processes and systems. There is much safety management information held in worker knowledge. Identification of critical steps is best completed by simply asking the workers what the critical steps are, and what should be done to defend against potential failure in those places within the process.

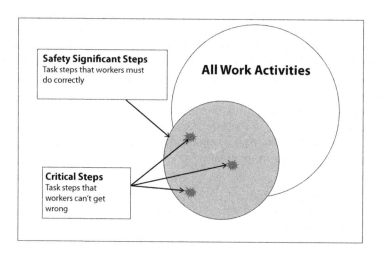

Figure 6.1 Critical Steps Imagined and Prioritized

Case Study
A Crashing Limb: Thinking about where Failure will Happen

Before you read this case study, note that we have talked quite a lot about the importance of how your organization reacts to failure. This case study helps us understand the way the people involved in the failure respond as well. These three questions should help set the stage for your thinking about this story of a failure.

- How much does retrospective thinking influence the workers involved?
- Do employees accept blame in order to save time and organizational discomfort?
- Is it possible to miss complex relationships because of an immediate focus on misjudgment of the worker?

I fell in love with my house the moment I saw it. I had always had the desire to live in a bungalow-styled home, and this one was to be my first "home." I was out of the renter lifestyle, and was finally a homeowner. It wasn't really a huge house, but it was mine. It was lined up on a street with other houses similar in style and size. I loved this old neighborhood. It was a block of charming, well-kept homes surrounded by an old giant Douglas fir, big leaf maple, and western red cedar trees. The house had two bedrooms, one bathroom, a spacious living room, and a typical "A" frame roof, and—my favorite feature left by the former owner—beautiful copper rain gutters.

I had been in my bungalow for two years. I spent any extra dollar I had on improvements. I hired a gardener to come every month and maintain the yard. I repainted the outside trim. I made sure that it was a "home," but I also kept pride in making sure it was one of the nicest places in the neighborhood. My house was even featured in an article in the paper on the revitalization of bungalow houses in the city.

On the night of the accident, the 10:00 pm news predicted a small storm passing through the area overnight. I locked up the house, turned off the television, and crawled into bed. I could hear the wind picking up, but really didn't pay it much thought. It was mid-spring, and constant wind was pretty typical at this time of year.

I was awakened by what sounded like a missile being launched into the house. My alarm clock noted it was 3:17 am. My heart was pounding, and I was groggy, but adrenaline automatically took over as I rushed into the living room. When I ran into the room, I saw that the giant top of a tree had pierced its way right through the roof. Damage was everywhere. My television was a broken, black plastic electrical mess. I couldn't see my couch through all of the pine limbs and needles. My laptop was split in two. There was crap everywhere. The wind and rain rushed into my home, making this horrific scene cold and wet. My beautiful home was littered with pink insulation, broken wood planks, tarpaper, and pine limbs; there were needles everywhere. To this day I can't tell you how a long piece of the copper rain gutter made it through the hole. The wall clock that somehow stayed on the wall read that it was only 3:20 am. All of this happened in minutes, but seemed like a slow motion movie that took hours.

A few hours later the storm passed and the sun rose. In the early morning light, I could see that my beautiful neighborhood looked like a war zone. There were tree limbs and debris littered all over the street and yards. Bicycles, trash cans, and planters were knocked all over the yards of my neighbor's homes. The bushes and early spring flowers literally looked as if they had been shredded. But my house was the only one that had sustained any substantial damage. I was obviously disappointed. I was thankful that nobody was injured, but angry. Why did this happen to my house? I maintained my house almost compulsively. I am a pretty nice guy. Why did this happen to me?

The meteorologist on the news the night before had said that it would be a small storm with a chance of showers. I found out from my neighbor that what we had experienced was a bizarre phenomenon called a "microburst." A microburst is a severe, localized wind blasting down from a thunderstorm. A dramatic wind shear is erratic and virtually unpredictable. A microburst only affects a small area—less than 2.5 square miles—and it happened at my house. The weatherman predicted a small storm and a deluge.

My gardener had been in the yard two days before, clearing away leaves and prepping the yard for the summer. He was up in the trees, cutting off some of the dead limbs. He informed me that some of the live limbs that scraped the roof couldn't be cut off until the summer or early fall, as such cuts could damage these 100-year-old trees. I wondered if he might have accidently cut some of the live limbs on the trees and hadn't told me. I didn't know if I should fire him, or even possibly sue him for negligence. He was the professional; he should have warned me that this could happen. I would have expected this to happen last winter, when the trees were covered in heavy snow and ice, but not now.

I had read an article in the paper several weeks previously that said our region was being infested with the mountain pine beetle. But the article said that the beetles were only attacking old or weakened trees. The beetles lay eggs under bark and then constrict the way the tree circulates sap, basically suffocating the tree. My trees, although old, were healthy and strong. The article stated that the unusual hot, dry summers and mild winters of the past few years had led to an unprecedented epidemic in the region. However, it also stated that the beetle was in the forest and had yet to move into the city. I had made sure to water my yard and trees. I began to think that the beetle infestation must have reached my neighborhood and affected the trees surrounding my house. The authorities should have given the public better warning that the infestation was going to be massive. My trees didn't display the "red tops" mentioned in the article. My trees looked pretty good—well, until one javelined into my living room.

In hindsight, I guess I should have paid more attention. I was well aware that old, giant trees surrounded the house, but they were so beautiful. I didn't think anything could happen. Some of these trees had been there for over 100 years. I should have been smarter and cut down all the trees that were next to the house. I should have spent more time looking at the trees, and making sure they were healthy. I should have talked to the gardener about cutting the limbs. I should have made sure the roof was in better shape, structurally sound. I certainly should have been aware that a mature tree weighed 40,000 lbs., and could cause $115,000 of damage.

RETROSPECT MAKES MISTAKES MORE SIGNIFICANT

It is easy to look backwards at an event and see exactly where mistakes in judgment were made. Every worker involved in a failure does exactly that. These workers replay events and decisions over and over in order to pinpoint where they went awry. What is happening is the process of making sense of the unexpected event. This process is important, but it also is meaningful to your understanding of Human Performance. The workers involved will often punish themselves more than the organization is going to punish them.

Why does this self-inflicted admission of guilt happen? Three reasons:

1. It is faster to admit a mistake and accept blame than it is to wait for the organization to *make* you admit a mistake and accept blame. Take your medicine early, and all of these complications go away quickly, and we can get back to normal.
2. If the worker takes all the responsibility, other workers will not be implicated in this event. The worker takes the bullet for the team in order to keep their friends out of all this trouble.
3. We are emotionally wired to blame ourselves for mistakes that, in retrospect, seem stupid.

Framing failure as a product of mistaken judgment is easy, but not accurate. The same decision that is seen as a mistake after the microburst windstorm was an example of wise stewardship of older trees in the event context. What changed was not the quality of the decision. What changed was the context of the event.

This case study is a compelling example of a complex failure. The criteria for a complex failure can be found in three characteristics:

1. The conditions in this story are all interdependently related to one another—trees, weather, bugs, gardeners, and roofs.
2. The environment in which the failure happened is so able to adapt that failure is very unusual. The system (in this case the tree) was stable for more than 40 years.
3. All the failure conditions in this story are connected together not by unusual reasons, but by remarkably normal reasons. All of these above mentioned conditions always existed together; in fact, trees and gardens and weather *have* to go together.

This failure, and the opportunity to see this failure through the eyes of the homeowner, seems both innocent and simple. This failure is anything but simple. This failure provides insight and motivation to look into the context of this story, even beyond the homeowner's impression.

7
Fundamentals Training: Introducing the "New View" to Your Old Crew

QUESTIONS ABOUT YOUR ORGANIZATION:

- How do your workers define safety and performance success?
- How do your managers define safety and performance success?
- Are they the same definition?

Performance certainly is best defined in your organization as the moments when what you expected to happen, happened. As simple as that definition is, training is particularly valuable to your organization's idea of expected performance. The only way your workers will know what you expect of them is if you tell them. If you don't tell them what you expect, your workers will go out and figure out the answer themselves.

Perhaps the most fundamental lesson in this book is this lesson: never assume that a group of managers or workers has a level of awareness about Human Performance unless you have talked with them about this topic before. An assumed understanding of these ideas is both offensive to the group and dangerous for your ability to create a successful training session.

With every group you will train, you have to start at the place where the group is at the time of your training. When you assume the group knows all the basics, that assumption will jump up and bite you every time. Do your homework, talk to your students, ask around your organization, and be sure you know at what level your training should start so you can help ensure success for the group.

The problem is that your excitement and maturity for these ideas begin to move quickly as you become more familiar with the new view of safety management. As you read more and more about this topic, you will want to move to the newest ideas and concepts. This desire to start at a level well beyond the basics is noble and predictable. You want your organization to benefit from the newest ideas. In my experience, when you jump to the advanced ideas you take the opportunity away from your learners to benefit from the progressive nature of these ideas. You are teaching your organization to run before you have allowed them to learn to walk.

I was asked to introduce Human Performance to a large group of executives just the other day. Not an unusual request, but this request was different. This request made me stop and think. In this case, the requestor asked me to present the Human Performance of 5 years ago, not the Human Performance information of today. That made me think. What this group needed was the chance to start at the beginning of the conversation, not to be dragged into the middle of a conversation. What a powerful lesson for me (and hopefully for you as well—I can almost guarantee you will overshoot your audience at one point or another in your training program).

Training is meaningless unless it has real value and purpose. We all deal every day with trying to make our safety training meaningful and valuable—something much more than just a "check the box" exercise. What has helped me in my career is one simple word with enormous power. That word is simply "intention." Our job is to give our workers training that has a direct and intentional outcome.

Human Performance depends so much on the successful and effective introduction of these concepts. In many ways, you will have only one chance to make this impression about this new view of safety thinking. This work is indispensable. This training cannot be an afterthought. Leave nothing to chance—create training that is purposeful and intentional.

DO YOU REALLY NEED *MORE* TRAINING?

It does seem like there is a lot of safety training. Your workers are acutely aware of how much training is required of them. Managers and supervisors are even more aware of the time, energy, and resources that are used on your training efforts. Training is often used as a quick and easy fix for a safety problem. Something bad happens, and the next thing you know there is a training session to tell your workers not to do that bad thing ever again. Your organization probably has enough of that type of training.

Human Performance needs time, effort, and training resources in order to be effective.

I would set out the expectation early in your change management strategy that training is vital and will not be compromised. It is predictable that your management team will not want to take an entire day to participate in Fundamentals of Human Performance training. These are busy people, and they rarely take a whole day to do any activity. However, don't shortchange this group by taking away time for these people to process information. They need the chance to talk about these ideas as much as (if not more than) any worker.

Building knowledge in your organization is about much more than training. Yet, training matters. In my organization, we learned early that the amount of time we spent creating the opportunity to teach these theories and ideas was directly related to the amount of success that we had in changing the safety performance. That being said, it is important to emphasize the difference between talking about safety and Human Performance, and actually practicing safety and Human Performance.

Talking about safety does not create safer organizations. Workers in the field operating safely in a varying and unpredictable world are actually being safe. Never confuse the safety training with safety. Training is essential, but it is not nearly as important as workers detecting and correcting potential failures at their worksites. Remember this: workers keep your organization safe.

Human Performance is a safety management philosophy, not a program. The training you will provide your workers with will not really have an ending. There is no 5 part model or list of 17 things you should do to complete a new view of safety. The class that you will present to your workers and manage is going to be much more like a conversation than a training class. Don't despair—this seems to be the most effective way to talk about these ideas with real workers. Don't try to have an ending, try to have an effective conversation.

You're still left with the substantial challenge of how to get your workers and managers to understand this new language, these new concepts. How do you get them to grasp these new ways of observing and understanding failure? No matter how much time you have or don't have, the answer to this question is going to involve some type of worker training.

Safety training can be difficult. Much of safety training is time spent telling workers what not to do in the field. You know this type of training: don't get hurt, don't take shortcuts, and don't screw up. Sadly, much of the field training across the country is not much better than this example. It is up to you to make this training as highly valuable as you can.

Training can be valuable and effective. It's not easy to make good training, but it is vital for developing a basis of thought for using Human Performance. Don't assume your workers or managers will magically receive this information. You owe your managers and workers the best, clearest explanation of the concepts of Human Performance.

TRAINING'S 5 INTENTIONS

I have done a lot of Human Performance training—hundreds of 8 hour classes a year, which is a lot of training.

What I have learned through the luxury of experience is that training is better when you do it deliberately, intentionally. The training that you will do will need to be as effective as you can make it. Your training will need to add value, be interesting and challenging, and be memorable. That is a lot of pressure on you to deliver, and you can do it. Here are 5 things I know about effective training. These are the 5 goals of intentional training. Knowing these 5 keys will help you formulate your training and tailor it to your specific audience.

Training Builds Foundations

In Human Performance training *everything* boils down to building a foundation for your managers and workers to think about and act on these new ideas. It is not that these ideas are difficult. The ideas are quite simple—in concept at least. In fact, the ideas we have been talking about are a method to reapply a common sense, systems approach back into our workplaces. The ideas may not seem difficult, but these ideas are different—very different—and that difference means that old ideas must move away so that the new ideas have a space in the worker's mind.

The change of mind takes time and a bit of effort to digest operationally. What is most difficult is not acquiring the new knowledge, but, in fact, letting go of the old knowledge. This training must start at the beginning and move towards the new outcome. You can't start in the middle of this discussion and expect your audience to be in the same place. You have to take the group from where they are to where you want the group to be. Knowing the importance of the foundation, the place where these new ideas will be supported and reinforced, is extremely relevant to your organization's success.

Tell your students that you will be building this foundation together throughout the training day, and that building this foundation is a process. You will start with some basic concepts, and by the end of your training time you will have moved to using these basic concepts in some "real world" situations. Bringing your learners in on your goal for the day helps your students better understand what it is you are trying to do, and makes them partially responsible for a successful training day. I should also add that I often tell the class that this will be the best training they have ever had—it doesn't hurt to plant little seeds of success in their heads every now and then.

Training Creates Dialogue

A very effective way to create change in your organization is through conversation. Nicely helping people know that they don't know is a powerful way to start this conversation. Knowing this gives your training a whole new objective. If conversation creates change in any organization, then your training should be an opportunity to create conversation. Don't underestimate the power of this training intention. I am convinced that this truly is the most powerful tool in the entire change management and training toolbox.

Your job is to create conversations—conversations in your training rooms, but also conversations in your break rooms, hallways, and parking lots. What your training must do is create a space in your workplace to have a conversation about how we define, process, manage, and understand safety. Don't hesitate to allow your students a chance to talk with each other in small group discussion; don't be shy in allowing your class to talk to each other in the large group. Above all, create a chance to allow a space for your students to talk to you and to each other.

Build in places and times for your workers to talk to you and to talk to each other. Have discussions. Encourage discussion and questions. If a student pushes back on these ideas, then it demonstrates a good thing: it means that they are really processing the ideas. Not everyone will agree with you, and that is also positive. You are creating a place for dialogue, both internal and external, and dialogue is a vital part of the process.

Training Prioritizes Ideas

Human Performance class, like other concept based classes, does not have an end product. You will not finish the class with 5 things to keep a worker safe on the job site. There are no easy answers for safety. There are no easy answers for changing the fundamental management philosophy for safety management.

Because there is no clear ending to this training, or even this discussion, it is important to imagine what you want the learners to do differently when the training is completed. Our goal was really pretty simple, yet pretty profound: we wanted people to leave at the end of class with more questions about how we define, manage, and improve safety at our facility. If the students left class with more questions, then this would mean that we had introduced some new thoughts about the topic into the minds of the students.

Now this process gets tough. One thing your training will do is to force you to learn these ideas in depth, and at a real practical level. You will have to decide where you stand on some pretty serious concepts. By delivering Human Performance training (or at least assisting in the process—and you should be in one of those two places during your training) you will hone your skills to an unbelievable level. It is said that the best way to learn about a subject is to teach that subject. The harder the questions you receive from the students as the instructor, the better you will become at understanding the ideas. Better understanding of the ideas is the tool that you will use to determine which ideas and concepts are best to emphasize for your organization.

Better yet, this is the chance for your students to make some decisions about the importance and significance of the ideas for your operations. Let me give you an example. Ask yourself this question about your organization:

Is there a difference between a process step that you can't get wrong, and a process step that you must get right?

This question (a pretty reasonable question about critical steps in a high risk process) causes a learner to prioritize their thinking about safety decisions and organizational behaviors. The reason this question works well in forming ideas about Human Performance is that the question is difficult, and forces the learner to process a lot of operational information in ways that they have never been asked to process that information before. Helping your students determine what safety means to them personally and to the organization as a unit is the first step in building and prioritizing this information. This entire process is the vital first step in creating attitude change—the next intention.

Training Forms Attitudes

The whole purpose of this training activity is to help form attitudes among your workers and managers. Human Performance ideas, the new view, need to engage the hearts and minds of your organization. You want to create a sense of passion around these ideas. Your job is to help your students form new attitudes about safety.

Attitudes are the products of your workers' values; values are the things that workers personally believe are important to them about their work environment. You can tell a lot about the values of workers in an organization by how the workers behave. Values produce attitudes and attitudes produce behaviors. Simply asking workers to behave differently does not have much of a long-term effect on their values or attitudes. If you go to a worksite and observe workers not wearing hard hats, for example, you can almost immediately draw the conclusion that the problem is not in the workers' behavior (although that is what most would assume), the problem is more about the workers' attitudes. Attitude is primary to behavior, not the other way around.

All of these ideas around attitudes are crucial to the success of any safety program. By building the Human Performance message over time, starting at the beginning and moving through the process, you are more likely to build a message that is deeper than the normal behavior change message. You will be building a message that helps your workers change the way they think about safety.

> If you want to change your organization's safety program—change the way your workers think about your organization's safety performance.

Training Allows Practice

Ideas only matter if you can use the ideas in real life. You know that *talking* about safety is not safety. The same idea is true for Human Performance—talking about Human Performance is not Human Performance. Your workers will want to practice these ideas in a way that is meaningful, interesting, and effective. Case studies, learning activities, role playing, and video events allow your students to take all of the theoretical information you have given them and translate this information into some application. Human Performance training certainly needs this opportunity to practice ideas in real life.

I would be remiss if I didn't discuss some of the learning activities I use to create dialogue in training. I play games at the beginning of most classes— games that are rife with error, confusion, and emerging solutions. I play these games to signal that this class will be a bit different. I also play these games to create a discussion that travels throughout the entire class. Games also allow learners to practice skills in a safe and informative way.

WHAT'S IN IT FOR YOUR ORGANIZATION?

Ultimately, the goal of your Human Performance training is threefold:

1. You want to create a vocabulary in your organization about Human Performance.
2. You want to prepare your workers and managers to see failure differently, and, therefore, see safety differently as well.
3. You want to motivate your organization towards a new approach to safety. Making your organization a safer place to work.

All three of these goals are vital to your success. You will not succeed if you don't create a language you can all use to discuss these ideas. Words like "error," "latent condition," "precursors" are all meaningful words, but these words become more meaningful when you help create unified meanings— specific meanings for these ideas.

When you tell your organization that the way the organization reacts to failure is a conscious choice and that you are going to try to change that choice, this idea can be quite controversial in the beginning. You need time to explain these ideas, and your organization needs time to think about the explanation you give them.

Finally, telling your organization that you are going to try this new approach allows you the opportunity to ask your workers and managers to help make this change successful. Ask for help in making this change, and then watch how asking people to help you changes the commitment levels throughout your organization.

BUILD KNOWLEDGE ONE IDEA AT A TIME

These new ideas are not difficult concepts. In fact, these ideas are often more akin to common sense than they are related to some earth-shattering new idea. The basic ideas of Human Performance are quite simple; take a look at this partial list:

- People are fallible.
- Blame is not a valuable way to understand failure.
- Accountability for safety is clear and moves upward in the organization.
- Organizational systems and processes drive behaviors.
- The seeds of all future accidents are planted today.
- Everything an organization needs to have a failure already exists in the organization's systems, processes, and work environments.

All of these ideas are important in and of themselves, but they are more important when you help learners understand that these ideas make even more sense when you look at the list as a hierarchy. See if this list builds logically upon itself. In other words, see if this numbered list makes more sense:

1. People are fallible.
2. Blame is not a valuable way to understand failure.
3. Accountability for safety is clear and moves upward in the organization.
4. Organizational systems and processes drive behaviors.
5. The seeds of all future accidents are planted today.
6. Everything an organization needs to have a failure already exists in the organization's systems, processes, and work environments.

It is impossible to have a discussion with a group of well-intentioned workers about system influence if you don't start that discussion by thinking about worker error. If you don't allow your learners to think differently about error, the idea that an accident is the result of an organizational failure will be really difficult, because the class will want to argue that the worker was the cause of the accident.

Cause becomes more powerful if you don't understand the idea of error. Error lets you move a class from "you can't fix stupid" to the idea that "everyone makes mistakes." Building that Human Performance bridge through conversation is not just something nice to do—the conversation is vital in creating new ways to see old ideas.

WHAT YOU WANT TO DO—YOUR OBJECTIVE

The purpose of fundamentals training is to start building basic knowledge. Start with a solid instructional objective. What are the learners to know at the end of this discussion?

Here is our course objective for the fundamentals course:

"At our organization, accountability for safety is a trade-off. Managers must be able to create an environment where workers can make good decisions by:

• Recognizing the role of human fallibility in Human Performance.

• Identifying how organizational systems influence worker performance.

• Embracing the role of the leader to manage organizational systems and positively influence human behavior."

Pretty simple. Not terribly advanced ideas, not too abrupt a change, and yet this objective statement is not to be underestimated. There is much context contained in these 6 lines of text. This objective statement opens a dialogue around accountability, fallibility, and the parallel management of both behavior, and systems and processes. The statement introduces these ideas in a way that the class can continually build on throughout the day.

This objective statement works well with my population. The objective allows for several discussions early on in the class. We talk about accountability. We talk about the role of Management. We build a case for further discussions on human fallibility, system influence, and the parallel path of managing systems and behaviors at the same time. This statement covers a lot of ground and creates a set of conversation guides for the rest of the class. Your objective statement will be different; your statement *should* be different to reflect the needs of your organization.

TRAINING: WHO AND WHEN

Most importantly, start with your management team. In order to successfully implement your entire program, you simply have to provide a basis of management knowledge. This group is busy and will want a shorter version of the training. Don't shorten your management training. If anything, managers need more time to discuss and strategize these ideas. If your managers don't know how and why their behavior matters, you cannot expect any behavior except the behavior you have always observed. Start with this team early—you will be glad you did.

The next population that should have special attention early on is your field supervisors. These are the ground level experts on worker performance, work management, safety, and production. This is when your training becomes a little less theoretical and a lot more practical. Now you get to have conversations with field experts who will help provide an honest or practical approach to what will work and what may not work as well in your organization.

Think about this class as an opportunity to discuss the workplace in the future tense—discuss the workplace you want to have, not the workplace you currently have. Participants will want to complain about how bad the current world is. There is no doubt that when you compare the old ways to manage safety with the new ways to manage safety there will be a notable difference. Allow people to feel the way they feel about the current state of affairs, but work hard to keep the conversation in the future state. Talk about how work will be, not how work is,

Make certain your training has an arc to the structure and development of the class. The class should start at the beginning, start with the basic ideas. The training should build to a more and more detailed understanding of the theories, moving to a conclusion that tells the workers and managers what should happen next. Think of it like a hill in the middle of a grass field. Start on one side of the field, climb the hill, come back down, and end on the other side of the field. That is how your class should look and feel. Start easy, build on more complex ideas, and end in another place.

Search for three or four Human Performance principles that will make up your program for your organization. Use these principles to guide the development arc for your class. Pick these principles remembering that they will guide all the developmental work your organization will do to both introduce and manage this program. Our principles start with human fallibility, move through system influence, and end with parallel management of systems and behaviors. These principles help create that instructional arc, drive good conversation, and end with the challenge to leadership to manage differently (and more effectively).

THE LAST WORD ON TRAINING

This is your education tool, your marketing tool, your change management tool, your worker engagement tool, and your management guidance tool all in one. My advice is not to underestimate the importance of this training activity. Spend time and effort to ensure that your training is as complete, effective, and entertaining as it needs to be to create buy in and support for your program.

Look for programs everywhere…in other companies, other professions, other disciplines; look for any programs you think will be effective in facilitating change at your facility. You can do this and do it well. It is worth the effort.

Case Study
How to Win Friends and
Influence Workers

One of the challenges in every training class is getting started. This case study focuses on one method for starting a Human Performance class, with a highly interactive, dynamic learning activity. Try this activity with a group of trusted associates the first time, then move confidently through your organization spreading the word. Before you read this case study, ask yourself these questions about the role training for fundamentals will play in your organization:

- How does your organization manage the communication of new safety programs?
- What has been the most successful safety program for your organization?
- What was the best safety training you have ever had, and why?

Human Performance training is not about giving out information. There are many excellent books (most of them from Ashgate) that will keep a student of the new view busy for many hours. These books deliver exceptionally thoughtful ideas on how the past has directed our understanding of safety, what is currently happening in the safety world, and what the future holds in the new view of Human Performance management.

Receiving information doesn't seem to be enough for a room full of workers who have seen it all before, and are not about to believe *anything* based upon what the theory dictates. This room of workers does not need to be given information. This room full of workers doesn't know it, but what they actually want is to question the very foundation of what they think safety means. Sounds like fun, doesn't it?

NOBODY LIKES MANDATORY SAFETY TRAINING

Looking around this room you see a bunch of pretty hard-core looking workers in the back row, the first row to fill up in every class. These folks are quiet. They are sitting behind a table looking up at the instructor as if to say, "I dare you to teach me anything today." As time gets closer to the beginning of the class, the room starts to fill up. There are workers from every level of the organization.

As the workers arrive, the instructor moves around the room talking to students and asking questions. The questions seem a bit different from normal pre-class safety discussions. The instructor asks one of the harder looking, "back row" people if they are excited to be in class this morning. The instructor even asks if the participants went to bed early to make sure they were rested and ready for some training.

As soon as the class kicks off it takes a turn. There are not a lot of introductions; in fact, there are no introductions at all. Instead, the instructor says the following: "I will make you three promises about today's training: 1. At no time today will you have to touch each other—pause—unless you want to (wink, wink). 2. After speaking with your managers, we have reduced the amount of public singing to a more professional level. 3. This will be the best class you have ever taken; in fact, this will be the best mandatory safety training ever held on earth."

The room gets really quiet and the workers start looking at each other, and seemingly asking, "what is this class going to be like this morning?"

The instructor then tells the class that the reason the instructor said all that stuff about touching will become much clearer right about now. The instructor asks all the workers in the room to stand up and form a circle around the room. Twenty workers get up slowly and start to get into a circle. Several comments are made about not touching. The workers begin to think that this class might be a little different from the HAZCOM class that was held last month.

As the workers stand in the circle, the instructor begins his explanation:

"This is a pre-job briefing, and we are about to perform a task. This is a crucial task, so I need you to listen carefully to this pre-job briefing. It could get a bit complicated. For our task for this morning, we are going to count from one to 27 using ascending, numeric, sequential order—from one to 27. Let's try it. I will be one, you will be two, and you will be three. Everyone got the task? No tricky stuff—only whole numbers from one to 27. That is the task. Everyone got it? Give me some feedback. Oh, there are a couple of restrictions that I should tell you about. The first restriction comes from corporate. Corporate wants us to do this task as efficiently as possible. Time is money. No wasting time, no messing around; we are to get from one to 27 as fast as possible, and they are watching us. The second restriction comes from OSHA, and it is complicated and kind of hard to understand. I am not sure why we have to follow this rule, but I know that this rule is tremendously important, and they are serious about it. The OSHA rule says that even though we are to count from one to 27 in ascending, numeric, sequential order, we can never have a pattern of counters. The numbers have to be in order, but we cannot have a pattern of people counting. We can't go every other person, which would be bad. We can't go across the circle or go around the circle. We can have no patterns of people counting. The last rule is a local rule that seems to be pretty critical to the customer. The customer has stated that at no time can we have two, or more, people speaking at the same time. If that happens, that error is non-recoverable, and means we have to start again at one and go to 27. So, everybody got the rules? Let's get started: ready, set...go!"

And the circle of workers standing in the classroom slowly start to count. The class makes it all the way to three before two people call out "four" at the same time. The workers have to start again from the beginning. This time, the class progresses to 5 before three people call "6" at the same time. The workers start for a third time, but this time three people say "one" at the same time.

An individual in the class mentions that maybe they should individually raise their hands before they actually say the number. No other class members say anything, but slowly this idea starts to take off. The class progresses all the way to 12 before two people say "thirteen" at the same time. However, this time the class starts over immediately, each member of the class raising their hands and counting. Before you know it, the class has gone from one to 27 efficiently, without a pattern, and one at a time. The class has been successful. The task is done. Everyone claps and cheers as if the class has accomplished the most important task ever. Remember, these are pretty hardened workers who have been through a lot of this type of training, and almost never clap and cheer at work.

Then the instructor asks a question: "What just happened? Why would we play this game this morning? This game could be the most valuable thing we do all day, and I am curious as to why you think I had you play this game."

The workers start talking about communication and teamwork. A couple of people mention that they had to be careful and extra observant in order to get this task accomplished. Then the instructor asks the workers what they think about the pre-task instructions. The instructor wants to know if the instructions were clear, complete, effective, or even helpful. Were these instructions effective in helping to get the task accomplished? What type of instructions would you call these? Have you ever had task instructions like these in the field?

Soon, the class starts talking about how these instructions were given to tell the workers what *not* to do, but did not tell the workers what *to* do. The instructions set up the workers to fail, and the pre-job briefing did not help make the work more successful. These instructions were given to create compliance for the task, not to create mission success.

Then the instructor asks this question: "How important were your failures? You had four cycles of failure. Did these failures mean anything? Were these failures important?" One worker steps up and says that the failures were extremely significant because the failures allowed the group to learn. The instructor follows up this comment by asking, "what did the failures teach you?"

Now the class discussion. Twenty people still standing in a circle in the middle of a training room take an intriguing turn. These same workers who had before the class began been "less than interested" in what was going to happen that day start to discuss the idea that the mistakes immediately told the workers where the work instructions were weak and prone to failure. This discussion leads to much talk about how beneficial it was to estimate where the failures were likely to happen. The workers realize that if you know where a system is going to go wrong you can protect against that predictable failure. The workers are now interested.

The next question the instructor asks is how the class felt about their solution. How did the "raise your hand before you counted out loud solution" work for you? The workers decide the solution worked well. The workers seem to like the solution. For a low consequence game like this, the solution was perfect—not too complicated or time-consuming. The instructor then asks, "was the solution a process improvement, or was the solution a 'short cut?' "

That question causes a moment of silence among the workers, and soon a discussion starts about what the difference is between a short cut and a process improvement. Many workers feel that their solution was not a short cut—the solution was the only way the group could accomplish the task without failure. When pushed a little by the instructor, the group does admit that they added some detail to the work procedures and restrictions. The spirit of the procedure was there, the workers simply changed the process in order to accomplish the task successfully. The workers decide that the only difference between the two options is the outcome. Outcomes define the difference between a good idea and a bad idea.

The workers are then asked if they *chose* to follow the solution idea to raise your hand before you counted out loud. Was it a choice to decide to raise your hand? After much discussion, the workers decide that it wasn't quite a choice; instead it was what the group was doing, and so they all did it without choosing. The act of raising your hand before counting was the way the work was being done, not a behavioral choice. It becomes abundantly clear that the solution that was introduced to the group did not get introduced as a solution, but that the solution *emerged as an outcome of the group and the problem*; in short, the emergent solution was a product of the work context that was present when the group did the work activity.

And with that, the instructor tells the class they can sit back in their chairs. In the space of about 20 minutes, the instructor has set a tone that told the workers that this class would be different to all the other safety classes they had ever had before; and they have played a game that at first seemed stupid and complicated, but soon became a problem to be solved by the group, and then opened up the room for a discussion—not led by the instructor, but listened to—about things like the power of the system to create behaviors, the importance of context, the similarities between process improvements and "work around," the emergent nature of complex, adaptive solutions, and the purpose of procedures. It is as if the whole class has been previewed, all while the workers have stood in a circle, looking at each other.

And then the class starts—or had it actually started the moment the workers started to come into the classroom and sit down?

CAN MANAGERS BENEFIT FROM LEARNING ACTIVITIES?

So, you are telling me that this game is great at the worker level but may not be as effective for managers? After all, managers are busy people, and they are a little too important to play games. I would counter that argument by saying this: "Your managers deserve the opportunity to have this same learning experience." If you edit this out of their training for them, you are taking away the managers' ability to learn from the beginning. Anytime you start away from your fundamental knowledge, you risk bringing your management team into the middle of the conversation. You are starting them without the benefit of context.

LEARNING HAPPENS THROUGH QUESTIONS

This is how every fundamentals class I teach starts for both managers and workers. Every class gets to play the "circle game." That little dynamic learning activity is perfect for building trust, introducing concepts, encouraging discussion, and demonstrating emergent and adaptive problem solving. It is also fun to do and even more fun to watch.

Every class is a little different, and so every round of the circle game is a little different, but every group has an emergent solution that "appears" at some point in the process. Every group is extremely willing to adapt their notion of the process in order to complete the task. Every group wants to solve the problem.

DISCUSSION

Perhaps the most notable aspect of this game, however, is the dialogue it creates. The discussions are always rich, filled with context, and always illustrate the basic idea that workers are constantly detecting and correcting in real time in order to get work done. Quite honestly, I would be hard-pressed to find a better way to introduce so many concepts so early and so effectively in this class than this learning activity.

8
Starting the Journey
—The First Steps

"The journey of one thousand miles begins with the first step..."

QUESTIONS ABOUT YOU AND YOUR ORGANIZATION:

- Does your organization have a compelling reason to do things differently?
- Is there a need to move safety performance to the next level?
- Can you get management support for this change?
- Can you be in the position of teacher, coach, expert, and counselor?
- Are you committed and passionate about managing safety—*differently*?

Once you have made the decision that it is time to try a new way to manage and understand safety, quality, security, production, and systems, you must begin the task of getting this new view of the world to the people of your organization. Almost nothing about this task will be easy. It is my hope that I can share some of what I've learned, some of what I wish others had told me, and most of the things I would have done differently.

The first step on this journey of change has to be to both educate and inform your senior management team in order to get a commitment to support the change to the new philosophy of safety management. Take your managers to meetings where this view will be discussed. Introduce your managers to managers of other organizations that are on the same journey. Drop books off on your managers' desks. Look for articles and case studies that show the benefits of the new philosophy. Forward emails that you receive. Most importantly, create time and opportunity to have conversations with your management team. The best way to introduce these ideas is by telling stories, providing examples, offering data, and having conversations.

For a while, the responsibility for changing your organization's entire outlook on safety will fall solely on your shoulders. Look for or create a core group of people within your organization who can support you and the change that is about to happen. You will need to be constantly vigilant for like-minded peers and coworkers who "get it" early. These folks will be your messengers in the field and at the management level.

Don't wait too long for either permission or authority to lead this change. Be careful with this advice—I am not supporting your "overthrow of the castle." However, I am saying that if you wait for your senior managers to give you permission to change your safety program, you may be waiting a while. Not because your managers don't want the change, but more because your managers may not know that this kind of change exists as an option. The only real way you will ever know the limits of your authority is to exceed it. This may be a chance to see where the boundary of your limits is.

Look for work units or groups that share common values and goals with this new philosophy, and target these areas for initial responses and efforts. No need to select a population that is preloaded to resist this change. Find a target group or unit for this philosophical change that is "doable" and meaningful, and put this target group or unit in your sites.

I can assure you that you will have people in your organization who will adapt to these new ideas early and easily, and these people will be invaluable to your success. You will also have people who have been thinking this way for a long time, and will welcome the fact that the organization has finally caught up with the way they have been thinking. The support you get from these early adopters is helpful to you personally, as well as reinforcing to your new safety management view.

In reality, working from within your organization to facilitate change is all about fostering, building, and maintaining relationships. It is exactly the same way you have managed your safety program so far, the only difference is that you now are moving discussion away from ensuring compliance and fixing worker behavior, and toward learning from events, and monitoring system influences and factors.

Pick an area of your organization that you can use as your test area. Try to identify a group or unit of your organization that is proficient at change, or perhaps needs a lift, or is most ready for change for whatever reason. Many times during my journey I was introduced to divisions that were in terrible shape. These divisions were ready to try anything to get better. Once you have identified this test group, start their journey with a solid presentation of the fundamentals of Human Performance. You must help this group build a common understanding and vocabulary around the new view philosophy. In other words, you will have to educate this group on the fundamentals of Human Performance.

The initial training and introduction you give to any of your managers and coworkers are vital to your success. Don't think of Human Performance fundamentals training like you would think of standard safety training. This training is actually a guided dialogue. You are leading a discussion about this new way to view failure (and safety) around your workplace. Expect many questions; expect pushback. I think it is fair to expect some of your workers and managers to become a bit angry. Change is a scary concept for some people.

Remember, you are asking people to give up tools they believe have been incredibly effective throughout their entire careers. You are also asking managers and supervisors to potentially surrender control, to give up on ideas that they feel have worked for them. Often this need for control is expressed in terms of being able to hold workers accountable for their behaviors. You aren't asking these managers and supervisors to relinquish control. You are providing them with a whole new skill set. You are also not giving up on accountability. You are asking them to look at it in a bigger way. Trying out new ways to think about and manage safety may, at least initially, feel like surrendering control for this population.

When your managers balk at not being able to hold workers accountable after an event, what your managers are really telling you is that there are some fears and nervousness about being able to control workers. Unfortunately, worker control is a bit of a fantasy when you consider the ultimate fact that most workers make errors that cause failures—not intentional sabotage. It does very little good to punish a worker for doing something that they did not mean to do in the first place.

We are not introducing a way to reduce performance accountability. Human Performance asks one fundamental question about holding workers accountable: is it possible to hold your workers more accountable for safety than you do now? Human Performance talks about accountability more in terms of pulling it back up over the organizational systems and structures. Accountability is about the "account" of the event, the story of the failure. Culpability is about blame. Accountability for safety should be clear in your organization, and move *upward*.

Once you get some people up to speed on this new philosophy, identify these people as your field practitioner team. These people will be able to go to other groups and teams, and help those teams learn the new ideas. This core group is not set up to reinforce each other (they will be learning the ideas together), but to go into your organization to help set the stage for what is about to happen. This field practitioner group will also allow you to test ideas and plans against people who know what will work and not work in your organization.

You are changing your organization the way that every organization is changed—one person at a time. If you're lucky, you will change a set of people, then they will change another set of people, and before long you will start to see and, more importantly, hear the language of Human Performance being spoken at toolbox talks, pre-job briefings, and accident investigations. That is when you will know that your efforts are working. On that first day you hear the new language, you have my permission to buy yourself a beer or get a massage.

WHAT YOU CAN DO EARLY AND WHY

Smartly and strategically placed little steps can have a substantial impact and start change. Knowing what I now know, I can assure you that two of the most powerful small milestones with immense impact are:

1. Senior Management training and concept introduction
2. Senior Management *reinforcement* of the training, and concept introduction

The message here is simple. These are discussions and reminders you will have to have many times with your managers. Expect your management team to backslide; it is almost certain to happen if the event seems to be significant. The old "name, blame, shame, and retrain" model is incredibly strong. Managers will fall back to a position they know out of fear and confusion. Remember: change is sometimes scary, and old habits die hard.

Don't assume your managers magically know how to lead the change your organization is about to make in the way you manage safety. Your managers won't know how to make this change happen. You will have to have many conversations about how to think about making these new ideas have life for your organization. Don't underestimate how quickly this new way to manage safety is becoming a marketable advantage for many contracts. In the nuclear power industry, most major construction industries, the medical industry, and others, an active Human Performance program is a baseline requirement in order even to submit a competitive bid. This new safety philosophy is a competitive advantage. Remind your managers of this fact.

The other immediate change you can make, which I have found to be surprisingly easy and extremely effective, is to change the way your organization, and, therefore, your organization's management team, responds to failure events. You will first have to determine how your organization currently reacts. Dekker, in his book *The Field Guide to Understanding Human Error*, says it best, and I quote him every day: "If you want to understand failure, you must first understand how your organization reacts to failure." Cut this out and tape it to your computer monitor.

You will be surprised at the change in your organization the first time one of your managers asks in an incident review, "how did we set up a worker for this failure?" as opposed to the normal question, "what were you thinking when you stepped off that scaffolding?"

In the interest of being efficient, here is a list of what to prepare for to move your organization to a more progressive view of failure:

- **Read everything you can.** Know that you don't know. In the back of this book, I will put a partial reading list to make this even easier. Don't limit yourself to books: check out magazines, newspapers, and the Web.

Look for articles that demonstrate a more systemic approach to safety. Read with an eye for the new view, and you will find lots of stuff to share and learn from.

- **Find a person who has made this change (or is making this change) in their industry, and make a friend.** Do this early. Don't be afraid to call this person and ask questions. It is so comforting to have a mentor who can help you understand difficult concepts and situations.
- **Start looking at what type of training you should be doing.** How will you both educate and motivate your people in this topic area? Ask people to see what training they are using in their companies. Beg, borrow, and steal from others. Use the work that is being done by other organizations to build your program.
- **Fully gain management commitment.** As we discussed earlier, this is important to the longevity of your program and the success of your change. Don't expect managers to know anything; don't assume knowledge that is not there. Teach your managers about what you want to do differently, and then talk them into responding in a new way.
- **Regain management commitment.** And then do it all again. You will be the guiding light for management decisions made around safety and performance management. Check managers' attitudes and behaviors constantly. Allow managers to start with old ideas, and from there you can guide them to the new ideas, building a bridge from the old knowledge to the new knowledge. This will be your most repetitive task.
- **Identify your core group and opinion leaders.** Figure out who "gets it," and ask these people to be a part of the change you are about to facilitate. Build strong relationships with these people and provide sincere reinforcement. Allow them to be a part of the successes of the new view, and shield them from the failures that will inevitably happen.
- **Check for backslide activities.** And there will be lots of backsliding at first. Remember, the old view was strong and effective for a long time. It is not that the old view is bad, it is that this new philosophy is better and will take your organization's Human Performance to the next level.
- **Capitalize on worker involvement.** Involve your workers early in strategies, planning, and ownership for the success of your safety programs. Have safety committees. Include workers early and often. Then, most importantly, get out of the way.
- **Start change management strategy.** Make a plan, create a project, and use some pretty solid milestones, and then go out and follow your plan. Much of the change will happen ad hoc or by improvisation, but much of it will follow your strategic plan. Think about where you want this effort to take you (more near miss reporting, better post-job

reviews, more checklists, better worker involvement), and then build a backwards-looking plan to get you to those goals.

- **Begin building a unified vocabulary.** Use the language of the new philosophy in your reports, reviews, training, and at your management meetings. The best way to facilitate a new view of safety is to use the new view language in communication with each other at the job site.

- **Start Human Performance fundamentals training for your managers, supervisors, and workers.** The only way your workforce will understand the effort you are trying to start at your organization is if you tell them about it. You must think of how to communicate the concepts, ideas, and tools of Human Performance to all levels of your workforce.

- **Work on changing the way your organization does failure analysis and investigations.** You may get tired of hearing this message, but one of the most effective ways to communicate to your workforce that the organization is interested in managing safety differently is to change the way you investigate failure. First, it gives the organization an entirely new set of information about an event. Second, it communicates change through real actions—which is always much more effective than talk.

- **Look at work planning. Planning is always approximate and incomplete so in many ways planning is never done.** If a worker calls a Stop Work, your work planning has failed. How we plan work, how we predict potential error-likely situations within that work, and what tools and incentives we build in to the work planning all should function collectively to help workers perform safely and successfully. Planning lives at every level of your organization. Observe the planning activities and ask this question: "Are we setting our workers up to be successful?"

- **Look for "low hanging fruit."** Find the changes that are going to give you the most change with the least amount of resistance. Change your critique program, or change the way you do pre-job reviews, or start doing quick post-job reviews. Search for ways you can build success for your workers by using Human Performance tools and ideas to make decisive, quality change. The quickest and best change for us was to change our accident critique and investigation process. That was a minor change that had an immediate and positive impact on our organization.

- **Make your program visible and deliberate.** Tell people what you are doing and why you are doing it. The change will be much more successful if the workers are a part of it. The way to include workers is to talk to them about what you are doing, why it is important, what differences they can expect, and how you want them to behave differently.

- **Get to the field.** Make sure you are out, visible, and talking. Take part in pre- and post-jobs, be a part of how work is done, and actively seek "teachable moments." Spread the word about the new view. Talk about how the new view is giving the organization better information, and in turn allowing for more effective performance.
- **Try a "learning team" approach.** Don't be afraid to learn about every type of failure or near failure you have given to you. Put in place teams of people from every level of your organization to look at near misses, to address conflicts and problems, and to review new ideas, processes, or procedures. Tapping in to workers in a semi-formal "learning" setting will be refreshing and surprising. You will learn more operationally from these teams than you will ever learn individually.
- **Know that this takes time.** The proverbial "nothing happens overnight." You must be ready, and you must ready your management for the fact that this process will take some time to "grow its own legs." Have faith. You will be surprised at how much good you will be doing from the very first day of this new approach. The only problem is that it really is hard to measure all the failures that didn't happen because of your Human Performance improvement effort.
- **Help someone else.** Just as you learned from other organizations, others will want to learn from you. The benefit of helping other organizations is that you will be stunned at how much you will learn about your own organization.

YOUR CHALLENGE IS TO CHANGE BELIEFS

All of these suggestions are just simple suggestions. You will certainly have more to add to this list, and I hope you do. What is essential is that you move to this change with a sense of fearlessness. You are doing the right thing. This type of change is hard, and you will have periods where you may get angry and frustrated. Keep a sane head and your expectations low. Any improvement in the organization makes the next improvement a little bit easier.

Think in small steps. Think one person at a time. It does not sound efficient, but it is the only way to make change stick. You are helping to change people's beliefs about safety, and that is no easy task. When you think one person at a time, you create the opportunity for many small wins. In many ways, numerous small wins are more satisfying than the big wins.

Stay grounded. Start where the group is and help them move to where you feel they need to go. The group owns the problem and, therefore, must also own the solution—this isn't your solution for your organization—you are helping your organization find its own solution.

9
The Four Things that Matter

Through each chapter in this book, I have tried to give open, non-academic discussions on how to apply the "new view" of safety, performance, quality, and security program management. We have discussed tools and ideas that you can use immediately to increase your awareness and help have serious conversations with your management teams, and ways to build trust and get leading information from the plant floor.

Aside from the traditional metrics for success, what should you look for to tell you that your organization is learning about and from itself? To answer that, let me tell you a little story about a company that tried out this new Human Performance approach.

A large construction company was suffering a rather tough time, with lots of bad luck. In a short period of less than a year, this company had four failures. The downturn in fortune started with a vehicle accident in a company truck that killed one of the most loved old-timers. The accident left a scar that ran deeply through this company; not only had the company lost a friend, but also one of the safety leaders, and the best foreman they had. This company lost one of the clearest and most significant safety voices in a terrible accident. The loss of this friend was so profound that the company's workers requested that this worker's unit number be retired from service and painted on all of the company vehicles as a lasting tribute and an ever-present reminder of a lost fellow worker.

This event alone would have been sufficient to create change. But about a month later, this same company dispatched a truck to one of their construction sites. The truck arrived with no problems; the driver hopped out to speak to the site supervisor to coordinate the unloading process. While the driver was out of the truck, and with the truck parked on a slight grade, the brakes failed, and the truck rolled off the site. It barreled through a chain-link fence, across a street, and into an intersection. When the truck rolled into the traffic intersection, it hit a waiting vehicle stopped at the red light, killing the woman driving the vehicle.

The truck had been well maintained and well inspected. Unfortunately, the brakes still failed, and this terrible accident happened. Terrible accident number two was barely out of the shock and investigation phase when a third event took place.

This event was a combination of a production pressure, experimental techniques, contract scoping, and rotten luck. The construction company was building in a downtown urban setting; the building was moving along at a quick pace, on schedule, and on budget. However, there was tremendous pressure to deliver this project on time. In fact, there was considerable pressure from both the company owner, and the building owner and financing agency.

The pressure did not abate. The workers doubled their efforts, and the schedule was expanded to 7 days a week. Somehow, around a floor in the mid-teens, the project collapsed like a pancake. No workers were injured; in fact, the workers on the building were able to ride the collapsing building to the ground "like a surfboard."

Suddenly this company had suffered three serious failures in a row. Any one of these failures would have been enough to impact any business seriously. Three of these sent a serious message from the top to the bottom of the company.

I was asked to come and meet with this company. I spent a long weekend—every day, all day, for three days—with the owners and the senior managers of this company. We then followed these three days with a full day workshop with all the field supervisors. We then created a plan to monitor low level, close call, and good catch data. We worked on creating a post-job learning process that got excellent information, did not cost lots of time or money, and totally changed this company's event learning techniques.

All of these things made an enormous difference quickly. Remember, this company had had three incredibly significant, "culture changing" events. Getting them to change everything was easy. This was truly an organization that knew they had to change.

AND THEN IT HAPPENED...

The Occupational Safety and Health Administration (OSHA) featured them at the national conference as a company that exemplified how to change safety operations. This company had gone from the very bottom of the barrel of safety performance to the very top of the regulator's list of best practice performers.

One afternoon I got a call from these guys asking if they could use some of the materials we used to manage their safety performance change program. (I am a firm believer that information is to share. There is no value in our business in keeping a good idea for one's own benefit. If we can save a life on any job site, to me that is much more valuable than selling a set of slides on a Thursday afternoon. We all learn from each other, and that should never stop.) I said, "sure."

Here's what they told me that I told them. To be honest, I am not sure I said any of these things, but these following points are the "four principles" that this company used every day to go from worst to best.

COMPANIES THAT MANAGE SAFETY PERFORMANCE WELL CONSTANTLY WORK ON THESE FOUR THINGS

1. **We are fixated on where the next failure will happen.** Like all good organizations, good companies never want to be surprised by failure. These companies are constantly looking in the field for areas that are confusing, super-risky, or high pressure, and places that "just don't feel right." These companies know that they will not predict the next event, but they can predict environments where events and failures are most prone to happen. The trick is that when they find a place they believe warrants their attention, they give that job site their attention.

2. **We constantly strive to reduce complicated operations.** Good companies do complex work, it is just that they have learned how to do complex work in uncomplicated ways. There is no doubt that you could walk through your job sites and find terribly complicated processes, procedures, instructions, rules, and methods. You must ask yourself, "does this operational complication make work easier to do? Or is this complicated system serving some part of the organization other than the worker?" We found that we were writing our procedures and work control documents to avoid compliance failure, and not to contribute to the performance of the mission. Systems and operations usually start out with the worker in mind, but as time goes by your rules and processes drift towards information about maintaining compliance, not about maintaining production.

3. **We respond to low level signals seriously.** I will never forget what the vice president said to me about this third principle. He called me from his office. During our conversation, I could just imagine him sitting behind his desk, feet up on the edge of his file drawer, phone headset in his ear. "Todd, we learned that when a field supervisor calls and says we have a problem that could get bigger if we don't fix it, that is the moment to drop your pencil, get in the truck, and drive out to that site. That's our emergency. That's when we react emotionally. That is when we make the biggest difference. That is our job now: we go out there, and fix the problem." There is not much I can add to that, other than to say he is right.

4. **We respond to events deliberately.** Finally, when something fails— and something will fail, even the best companies have failure—good companies respond deliberately. The construction company made a

conscious decision that they would respond to failure deliberately. They don't get emotional; they don't go out to fix the worker. They don't enact immediate policy and rule change. They slow down and learn. Companies that are proficient at driving change in safety performance know that the only way that change can ever happen, the only way events are prevented, is through learning.

These four principles, wherever they came from, worked well on the heels of a significant safety culture shift. These same ideas will work for you. This message is also deliverable to workers, and to managers, owners, and customers, because this message is the right message.

But this message demands education and commitment from all levels of your organization. You must commit to these principles, and then hold yourselves accountable for doing these four things. Don't assume that by telling your organization these principles are indispensable you will ensure that your organization will run right out and change. Instead, teach your organization about how these principles will look, act, and feel in your company, and why. Don't assume that the same level of knowledge and maturity you are gaining will magically exist in everyone else. Old habits are hard to break. You will have fellow managers who will exhibit a high need to go out and fix workers who screw up. Fixing workers who screw up is acting emotionally, not acting deliberately.

Conclusion

Any attention paid towards your organization's system influences on Human Performance is a good thing for your organization. Simply by seeking to understand the story and context that surrounds a failure, you will see safety get better. You know that your organization is not perfect; you know that your processes, procedures, and the way your organization responds to events have a real opportunity to be better. New, increased attention will have an immediate benefit to your safety performance.

It is almost certain that after your research of these concepts and ideas you will have been left with more questions than answers. What should we do? How do we know we are doing the right things? What are the right things? Where do we start? Directing a new safety course correction is complicated, time-consuming, and, flatly, a lot of work.

The answer to all your questions is probably this: what works for your organization is going to be what your organization needs, and every organization needs something a little bit different. Remember: your job is to help change the way your organization learns about failure. When your organization begins to learn differently about failure, the organization will respond differently to failures, and that difference in most cases changes the way people in your organization communicate, understand, and correct the conditions that cause future problems, and events.

The ability to detect and correct all the conditions that normally exist as precursors to failure is not easy and, most importantly, not accidental. A pre-accident investigation is a deliberate and purposeful attempt before a failure to discover these normal conditions that could combine to set the stage for failure. What is important in the decision to be deliberate and purposeful is to change the way your organization gets to better, more reliable performance. Remember: simply asking workers to be more careful does not equate to more careful workers.

The other side of the coin is that everything that we have discussed in this book could be wrong, or maybe not wrong, just not quite right for your organization. Nothing will happen if you don't test these ideas.

Chances are, however, that these ideas are actually more of a way to reconnect both your managers and workers to some common sense ideas about performance. If performance is the idea of getting what you expect, Human Performance is a way to move your organization's systems so they set your workers up to be successful. As your systems stand now, your workers are finding processes and programs that create conflicts that must be solved in the field.

All of us want the same thing. We want every worker at every job to be healthy, happy, and safe. We want the singular best performance (whatever best performance may mean) for every worker in our organization. The problem is we know that dream is not possible, and so every single day we must think about new ways to help our workers avoid injuries, and to mitigate the consequences of failure.

Our journey is not over. The challenge is for you to think of new and effective ways to manage your company's systems and your workers' behaviors, in parallel, while not causing further complications or unintended consequences. We must keep thinking about the problem of managing systems and behaviors while not causing unintended consequences, thinking about how to measure the issues we discover, and know that the work is never done.

I challenge you to try new things, measure them somehow, and then repeat the cycle.

I am not planning on giving up. This work is too important. Too many people are counting on us; they don't even know they are counting on us, but they are counting on us.

I don't think you should give up either.

Basic Reading List
for Human Performance

START HERE

Dekker, S., *The Field Guide to Understanding Human Error*. Farnham. Ashgate Publishing

Gawande, A., *The Checklist Manifesto: How to Get Things Right*. New York. Metropolitan Books

Levitt, S. and Dubner, S., *Freakanomics*. New York. HarperCollins

Reason, J., *Managing the Risks of Organizational Accidents*. Farnham. Ashgate Publishing

THE NEXT LEVEL

Hollnagel, E., *Barriers and Accident Prevention*. Version 1. Vermont. Ashgate Publishing

Hollnagel, E., *The ETTO Principle: Efficiency-Thoroughness Trade-Off*. Farnham. Ashgate Publishing

Reason, J., *The Human Contribution*. Vermont. Ashgate Publishing

Weick, K. and Sutcliffe, K., *Managing the Unexpected: Resilient Performance in an Age of Uncertainty*. Second Edition. San Francisco. John Wiley & Sons

SENIOR CLASS LEVEL

Dekker, S., *Just Culture*. Farnham. Ashgate Publishing

Reason, J., *Human Error*. Cambridge. Cambridge University Press

Taleb, N., *The Black Swan: The Impact of the Highly Improbable*. New York. Random House Publishing

Wood, D., Dekker, S., Cook, R., Johannesen, L. and Sarter, N., *Behind Human Error*. Farnham. Ashgate Publishing

Index

CPSIA information can be obtained
at www.ICGtesting.com
Printed in the USA
BVOW04*1553141216
470818BV00005B/21/P